WELL PLAYED

Building Mathematical Thinking Through Number Games and Puzzles

GRADES K–2

Linda Dacey, Karen Gartland,
and Jayne Bamford Lynch

Foreword by Kassia Omohundro Wedekind

www.stenhouse.com

Portland, Maine

Stenhouse Publishers
www.stenhouse.com

Copyright © 2016 by Linda Dacey, Karen Gartland, and Jayne Bamford Lynch

All rights reserved. Except for the pages in the appendix, which may be photocopied for classroom use, no part of this publication may be reproduced or transmitted in any form or by any means, electronic or mechanical, including photocopy, or any information storage and retrieval system, without permission from the publisher.

Every effort has been made to contact copyright holders and students for permission to reproduce borrowed material. We regret any oversights that may have occurred and will be pleased to rectify them in subsequent reprints of the work.

Library of Congress Cataloging-in-Publication Data

Dacey, Linda Schulman, 1949-

Well played: building mathematical thinking through number games and puzzles, grades K–2 / Linda Dacey, Karen Gartland, and Jayne Bamford Lynch ; foreword by Kassia Omohundro Wedekind.

pages cm

Includes bibliographical references.

ISBN 978-1-62531-034-7 (pbk. : alk. paper) — ISBN 978-1-62531-077-4 (ebook) 1. Arithmetic—Study and teaching (Primary) 2. Counting. 3. Mathematics—Study and teaching (Primary) I. Gartland, Karen. II. Lynch, Jayne Bamford. III. Title.

QA135.6.D3325 2016

372.7'044—dc23

2015022556

Cover design by Alessandra S. Turati

Interior design and typesetting by Victory Productions, Inc.

Manufactured in the United States of America

PRINTED ON 30% PCW RECYCLED PAPER

22 21 20   9 8 7 6 5 4

Contents

Foreword by Kassia Omohundro Wedekind	vi
Acknowledgments	viii

Chapter 1: Introduction — 1

- Why This Book? — 2
- Is It a Game or a Puzzle or an Activity? — 3
- How Is This Book Organized? — 3

Chapter 2: Supporting Learning Through Games and Puzzles — 5

- Using Games and Puzzles in the Classroom — 6
- Setting Expectations and Sharing Responsibilities with Students — 7
- Assessing Learning and Setting Goals — 13
- Fostering Productive Discussions — 18
- Meeting Individual Differences — 19
- Organizing Students for Success — 19
- Organizing Materials for Success — 22
- Working with Families — 23
- Conclusion — 25

Chapter 3: Counting and Ordering — 27

- What's the Math? — 27
- Count 20 — 28
- Number Jigsaw — 32
- Nim — 35
- Mystery Number — 37
- Order Up — 41
- Online Games and Apps — 44

Chapter 4: Base Ten Numeration — 47

- What's the Math? — 47
- Win 1,000 — 48
- Number Sort — 52
- Go Number Fish — 57
- The Number Is/What Number Is? — 60
- Number Touch — 64
- Online Games and Apps — 67

Chapter 5: Addition — 69

- What's the Math? — 69
- Make a Pair — 70
- Equal Values — 73
- Triangle Totals — 77
- Yahoo! 100 — 81
- On Target — 84
- Online Games and Apps — 88

Chapter 6: Subtraction — 91

- What's the Math? — 91
- How Many Are in the Cup? — 92
- Move Along — 95
- Take the Numbers — 98
- Name That Number — 102
- Subtraction Tic-Tac-Toe — 105
- Online Games and Apps — 108

Chapter 7: Addition and Subtraction — 111

- What's the Math? — 111
- Word Problem Bingo — 112
- Four of a Kind — 116
- It's Greater — 119
- Meet the Rules — 122
- Make Sense — 125
- Online Games and Apps — 128

Appendix — A-1
Puzzle Answer Key — A-79
References — A-81

Games and Puzzles Listed in Alphabetical Order

	Description	Appendix
Count 20	28	A-4
Equal Values	73	A-35
Four of a Kind	116	A-68
Go Number Fish	57	A-21
How Many Are in the Cup?	92	A-48
It's Greater	119	A-71
Make a Pair	70	A-32
Make Sense	125	A-76
Meet the Rules	122	A-74
Move Along	95	A-50
Mystery Number	37	A-11
Name That Number	102	A-56
Nim	35	A-9
The Number Is/What Number Is?	60	A-25
Number Jigsaw	32	A-6
Number Sort	52	A-19
Number Touch	64	A-28
On Target	84	A-46
Order Up	41	A-13
Subtraction Tic-Tac-Toe	105	A-60
Take the Numbers	98	A-53
Triangle Totals	77	A-40
Win 1,000	48	A-17
Word Problem Bingo	112	A-64
Yahoo! 100	81	A-43

FOREWORD

When I first opened this book, I remember thinking, "How will this book about math games and puzzles be different from the countless other resources on the topic?" Only a few pages in, I realized that *Well Played* is very different. This is a book about math games and puzzles, but it is also a book about building communities of mathematicians who work together to problem solve, talk about math, and figure things out. This is a book full of thoughtful and well-chosen games and puzzles, but it is also a book that offers a lens into how we might include this kind of play in our own classrooms in ways that are deeply meaningful and engaging for our students. It is a book truly rooted in the realities and possibilities of the classroom, which is what makes it such a valuable resource for teachers.

Well Played includes games and puzzles that check all the boxes for inclusion in the classroom: simple to learn but mathematically meaningful, with variations that allow for differentiation within the classroom, and nuanced enough to allow students to expand upon and deepen thinking through multiple experiences with the game over time.

There are several parts of each game or puzzle's description to which readers will want to pay special attention. Each game or puzzle's "How It Looks in the Classroom" section offers a glimpse into a classroom in which the game or puzzle is being introduced or played. Going beyond "how to play the game," these sections offer teachers ideas for launching the game or puzzle in ways that encourage curiosity, collaboration, math talk, and productive struggle.

Each "What to Look For" section offers several questions for teachers to consider as they work with students. I can imagine keeping these questions on my clipboard as I observe and confer with students as a way of guiding my instructional moves and helping me take useful anecdotal notes on students' strategies and understandings.

Each game or puzzle includes several exit question choices to use for student reflection. These questions offer an opportunity for students to synthesize and solidify understandings, either in the form of productive group discussion or individual journaling. The exit questions also offer teachers the opportunity to assess students' understanding and use this knowledge to plan next steps for instruction. Carefully chosen student work examples provide a reference point for teachers as they analyze and respond to student thinking.

One of my favorite games in this book is "Yahoo! 100," from Chapter 5: Addition. Too often math games focused on computation emphasize rote procedural practice over sense-making and thoughtful use of strategies. This game is different. In "Yahoo! 100," students turn over cards, adding multiples of ten and single-digit numbers. The goal is to get the greatest number of cards with a sum less than or equal to 100. This game supports students' development of efficient mental math strategies for addition, but it also focuses on building number sense as students consider their sum's distance from 100 and whether it makes sense to flip another card. Additionally, the authors' purposeful decision to include only cards that are multiples of ten or single-digit numbers allows students to develop addition strategies based on an understanding of decomposing numbers by place value.

Well Played also acknowledges the powerful potential of online math play, offering suggestions for games, puzzles, and apps at the end of each chapter. Though many online games focus on skill-and-drill and fluency practice, the authors have chosen games and puzzles that emphasize problem solving and offer students opportunities to model with mathematics.

While this book will be a prized resource of classroom teachers, as a blend of professional development and practical resources, this book will also be valued by coaches as they work with teachers to deepen content understanding and guide instructional practices.

As I read this book I kept thinking, "I can't wait to see what students will do with this game!" and "I know a teacher who will love using this puzzle with her class!" My copy of this book will soon be full of sticky notes and dog-eared pages, and a list of people with whom I want to share it. I know teachers will love this book as much as I do—thank you Linda, Karen, and Jayne for writing it!

Kassia Omohundro Wedekind, author of *Math Exchanges: Guiding Young Mathematicians in Small-Group Meetings*

Acknowledgments

We are deeply indebted to the teachers and students who collaborated with us during the development of this project. We explored each game and puzzle within a classroom, and teacher and student insights permeate this book. We are particularly grateful to students for serving as our consumer experts. Their feedback helped us fine-tune our thinking and play more with ways to embed key mathematical ideas into our discussions of the games and puzzles.

We are grateful to everyone at Stenhouse, but most particularly Toby Gordon, who showed such early faith in us. Her words "Write about what matters most to you" gave us the freedom to explore, reflect, play, and puzzle. And then, of course, she gave us such valuable and timely feedback all along the way. We are also grateful to our outstanding outside reviewer, who probed our thinking with important insights and intriguing questions. Thank you also to Chris Downey, Elizabeth Tripp, and Jay Kilburn for their care and expertise. All of you added greatly to the quality of our work.

CHAPTER 1

Introduction

It is a happy talent to know how to play.
—Ralph Waldo Emerson

Our whole life is solving puzzles.
—Erno Rubik

Did you have a favorite game or puzzle as a child? Why did you like it? Looking back, what do you think you learned from it?

People have engaged in playing games and solving puzzles for thousands of years. Games and puzzles continue to provide important opportunities for children to experience playful learning. And, as the proliferation of online game playing and puzzle solving shows, these activities continue to capture our interest.

Games and puzzles based on logical thinking are often linked to mathematics. Out of school, they are considered recreational. In school, games and puzzles often provide opportunities for students to practice skills. We see their incorporation in math lessons more than in other subject areas. Within mathematics, they tend to focus on computation, with the goal of increasing fluency. Games and puzzles are included in most mathematics curriculum resources, and teachers might offer them as a choice at a center, as an independent activity during math workshop, or as a rotation during instructional time.

Why This Book?

So, with interest in games and puzzles fully established and lots of games available to teachers through online sources and curriculum materials, why did we want to write this book? As a way to begin to answer this question, we'd like to share something we witnessed in a first-grade classroom when observing students playing an addition game.

The students were playing a game the teacher called *Seven Sevens*. The game required players to take turns rolling two dice, adding the numbers shown, checking the sum with their partners, and coloring in a circle every time the sum was seven. The first player to get seven sums of seven won the game. The opportunity to play the game followed a mini-lesson on finding sums by counting on, preferably from the greater addend. The teacher asked these students to play one or two games in pairs and to remember to count on to find the sums. For example, if they rolled a 5 and a 3, the students were to count on three from five, thinking or saying, "Six, seven, eight," to identify the sum of eight. As the students left the rug area, the teacher reminded them to practice the skill of counting on while they were playing the game.

The teacher circulated as the students played to make sure the rules of the game were clear and to gather assessment data about how students were adding the numbers together. We watched as Parisa and Riley demonstrated a turn for her. Satisfied that they understood the game and the counting-on strategy, the teacher moved to the next pair of students. As soon as the teacher left, we overheard them decide that they weren't going to count on and proceeded to count all of the dots on both dice to find the sums. After observing a few rounds of such play, we were very interested in what was going to take place when the teacher circled back to check in with these two students. We were amazed to see that without exchanging a single word, the girls began to use the counting-on strategy while the teacher observed them.

We know it is impossible to monitor all students all of the time, and this teacher had checked in with these players twice. When we debriefed, we noted our strongly held belief that telling students which procedures to use rarely deepens their learning. We also talked about how clever these students were at controlling their learning and wondered if they might act differently if the game provided them with more strategic choices or conceptual challenge. Our conversation then moved quickly to a general discussion about games and puzzles and how we might increase their potential for deepening students' conceptual learning as well as their computational fluency. It was clear that we wanted to consider the use of games in primary classrooms more fully.

We began our work by identifying our favorite games and puzzles and asking teachers to do the same. This process led us to adapt some games and puzzles as well as create new ones. Then we thought about how to deepen students' learning through their exploration of these games and puzzles. We asked ourselves key questions, including:

- Why is this game or puzzle worth exploring?
- How could student-to-student math talk be increased?
- What might teachers notice as students played the game or solved the puzzle that would inform future instruction?
- What assessment tasks could reinforce student accountability?
- What task would provide an opportunity to extend students' thinking?

Is It a Game or a Puzzle or an Activity?

One of the surprises of this work was how murky the distinctions can be among games, puzzles, and tasks. Is a computer game that requires a player to find clues and the correct path to reach a certain goal a game or a series of puzzles? Are we playing a game when we solve a puzzle? Is pinning numbers in order along a clothesline a puzzle, a task, or a game? Do young children view every activity where they take turns and use manipulatives as a game? Koster suggests that "games are puzzles to solve, just like everything else we encounter in life" (2013, 34). Note that both games and puzzles

- involve sequencing and pattern recognition;
- require strategy; and
- offer competition against an opponent, or the clock, or your own abilities to reach a solution.

There are, of course, some differences. For example, puzzles can be lost only by giving up. We have identified the games and puzzles in this book as either one or the other, but we found the following criteria important to both

- It addresses important mathematical ideas.
- It is engaging.
- It offers a range of difficulty levels.
- It requires and stimulates mathematical insight.
- It supports the habits of mind essential for success with mathematics and real-world problem solving.

How Is This Book Organized?

Chapter 2 addresses instructional decisions related to games and puzzles in the classroom. Our goal is to support teacher orchestration of gaming and puzzling as well as assessment of student learning. We pay particular attention to helping students take responsibility for their roles as players and puzzlers. Kohlfeld (2009) also identifies the need to instruct students in how to choose partners, take turns, be patient, and lose. We would add to this list teaching students how to work cooperatively, discuss ideas, persevere, and win graciously. All of these ideas are considered in Chapter 2.

The next five chapters of this book focus on content-specific games and puzzles arranged by content focus: counting and ordering; base ten numeration; addition; subtraction; and addition and subtraction. There are five fully developed games or puzzles within each chapter as well as a section that suggests online games and puzzles (including apps) that are appropriate for the classroom. The online section is less detailed as such resources change frequently. Within this section, we identify those electronic resources that are free.

The discussion of each of the five games and puzzles is organized to address the goals we identified when we began this project. Along with the expected sections "Math Focus," "Materials Needed," and "Directions," each discussion includes a section called "How It Looks in the Classroom," which shares a brief classroom story from our field testing. "Tips from the Classroom" provides further ideas for supporting classroom implementation, some of which came from our student field testers. "What to Look For" identifies key ideas and misunderstandings that our experience suggests will be tapped, allowing you to collect data to inform instructional decisions and note student growth over time. The "Variations" section suggests ways to change the game or puzzle to better reach the range of learners in your class, to sustain its worthiness as student learning progresses, or to adapt it to better fit another grade level than indicated in the classroom story. Nearly all of the games and puzzles can be adapted to serve kindergarten through second-grade students, and the suggested variations will help you make such changes. The section "Exit Question Choices" provides some suggested questions you can pose to students after their experience with a game or puzzle. These tasks serve as a way for students to bring closure to the experience while demonstrating their content knowledge. Such questions reinforce students' accountability for their own learning. Students' responses can be oral or written, as appropriate, and can inform your instructional decision making.

As teachers, we recognize the value in partnering with students about their learning. The more we communicate to students the role that games and puzzles play in supporting their understanding of key mathematical concepts and the use of mathematics in the real world, the better. Recording sheets and exit questions allow students to share their strategies and knowledge and provide teachers with the opportunity to assess learning. At the end of each game or puzzle discussion, the "Extension" section, as you would expect, gives you an idea for extending the learning. As you gain familiarity, you, or your students, may create other examples of such questions and tasks.

No doubt you may be familiar with some of the games and puzzles included in this resource, especially those that are classics; you have your own favorites, too. Nonetheless, we are confident that you will appreciate the opportunity to think about the use of games and puzzles in the classroom and find new ways to make their exploration more effective for engaging students and deepening mathematical understanding.

CHAPTER 2

Supporting Learning Through Games and Puzzles

As students enter their first-grade classroom, they return the math games they brought home over the weekend. They carefully put the games back in their designated places in the classroom math game library. As she does every Monday morning, the teacher has written the names of the games that were taken home on a piece of chart paper that hangs next to the math library area. After returning a game, each student writes a brief comment or sketches a picture on a sticky note and places it under the name of the game he or she played. These comments and drawings will be shared at the morning meeting.

Under "Go Number Fish," Isabelle writes, I saw lots of ways to show numbers and I beat my twin brother.

Under "Make a Pair," Alex, who played with ten-frames, draws a happy face and writes, 6 + 4 = 10. *He later shares that this is now a "fast fact" for him because he just knows it.*

As simple as this scene may appear to an outside observer, teachers know that it reflects deep values related to students sharing responsibility for their learning. Teachers will also recognize the instilling of routines and setting of expectations that must precede such behavior. Thoughtful orchestration of mathematical games and puzzles goes well beyond offering opportunities to play and solve. We need to think about when and how to introduce games and puzzles and when to make them available for small-group or individual use. We must help students understand the purposes of and expectations for playing and solving and look for ways to share the responsibility of the learning process with them. We want to set and assess learning goals, support math talk, and meet the needs of individual learners while pursuing such activities. We should recognize ways to organize students and materials to support success, and we want to involve families and caregivers in the playing and solving.

Using Games and Puzzles in the Classroom

We do not believe that the instructional potential of games and puzzles is being realized in most classrooms. Too often we've seen them provided as activities with no follow-up or offered to students as choices after they have mastered the related mathematical content. We've seen that teachers who implement good teaching practices, such as asking significant debriefing questions after a problem-solving experience, often fail to utilize such practices with games and puzzles. Further, the games and puzzles used most frequently in classrooms tend to only develop procedural expertise, without attending to conceptual understanding. As a result, many students experience games or puzzles as fun activities or time fillers but do not consider them as essential to their learning or as an important part of a lesson for which they are accountable.

Many teachers provide a game or puzzle station as a component of a three-rotation lesson structure (small-group meeting with teacher, independent work, and game or puzzle). Some educators recommend an instructional cycle similar to that shown in Figure 2.1.

Time	Group A	Group B	Group C
15 minutes	Introductory activity	Introductory activity	Introductory activity
15 minutes	Small-group meeting with teacher	Game or puzzle	Independent work
15 minutes	Independent work	Small-group meeting with teacher	Game or puzzle
15 minutes	Game or puzzle	Independent work	Small-group meeting with teacher

Figure 2.1 Possible instructional cycle

While such a schedule can be a useful format, it is only one way to include games and puzzles in our classrooms and one way to meet individual needs. If we are to use games and puzzles to develop conceptual understanding as well as to build fluency, we must reevaluate how we are using them and be willing to make some changes in our classroom instruction.

⟩ The Purposes of Games Change over Time

The same game or puzzle can serve many different purposes, and those purposes change with increased exposure. As summarized in Figure 2.2, we view the *introduction* stage as an experience that exposes students to different ways of thinking and piques their interest. This introduction often may be the focus of the day's lesson. Students have several opportunities for follow-up play in teams, beginning within the initial lesson, during the *exploration* stage. During this stage, students are engaged in conversations with peers throughout the playing and solving. They may also be actively involved with large-group discussions that can occur at various points in the process. Through such conversations and discussions, conceptual understanding deepens, new ideas become clearer, and generalizations form. During the *variation* stage, changes in the game allow for greater challenge, and as a result, the interest level in the game and its appropriateness for learning are maintained for a longer period of time. Frequent play or puzzling may be at a practice level, rather than intended for the development of ideas. This *practice* stage supports automaticity when such reinforcement is needed and is preferred by students over the typical worksheet. At this stage, learners are more likely to be working alone or playing against a single opponent. Sometimes games continue to be played as favorites, long after they have met the goal of supporting the development of conceptual understanding or computational fluency. When this *recreation* stage is reached, we encourage you to make the game an indoor recess option or have students play it at home for enjoyment, allowing your class to investigate other mathematical concepts in the limited instructional time available.

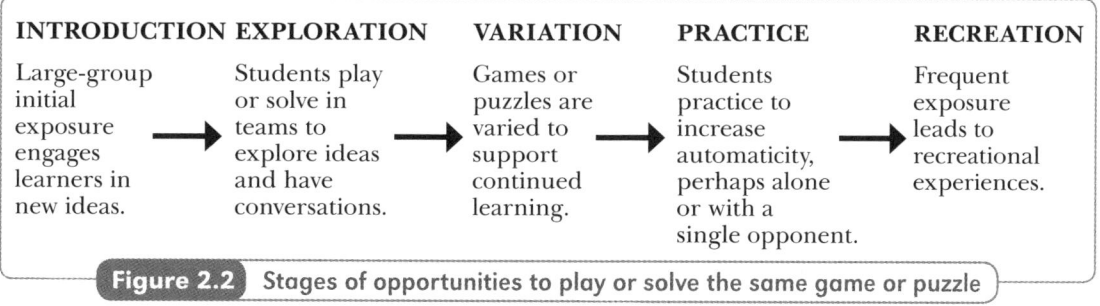

Figure 2.2 Stages of opportunities to play or solve the same game or puzzle

Setting Expectations and Sharing Responsibilities with Students

Some teachers view the setting of expectations and the sharing of responsibilities as separate topics, but we see them as intertwined. Students cannot be responsible without a clear understanding of what is expected of them. This is particularly important when the purposes of playing or solving include opportunities to develop conceptual understanding. Consider the following reflection from a second-grade teacher who understands that teaching students how to play a game is more than explaining directions and who believes strongly in holding students accountable for their learning.

TEACHER REFLECTION

On the first day of school this year, I noticed how energetic and social my students were. They always wanted to be talking to each other and be actively engaged in an activity. I was very tired that night, but also excited by this new group of young learners. I thought that playing math games would be particularly important this year as the games would engage them and give them the opportunity to talk.

I was quite surprised the next day when we played the game Mystery Number, *which required them to ask questions with yes-or-no answers to identify a two-digit number. They were all eager to ask a question, but their questions were all the same. Is it twenty-five? Is it fifty-two? Is it ninety-nine? I was modeling how to keep track of what they learned from the answers to the questions, and I crossed off each number they guessed on the hundreds chart that I had placed under the document camera. Then Carlos asked, "Is it greater than sixty?" I responded with "no" and then I crossed out the numbers 61–100.*

I was sure that students would notice the effectiveness of this question over the ones they had been asking, but the very next question was, "Is it twenty-two?"

I crossed out the 22 and then said, "Who can ask a question like the one Carlos asked?"

Dani immediately raised her hand and said, "I don't know what Carlos asked, but is it twenty-three?" When I asked if anyone remembered what Carlos had said, only five students raised their hands, and one of them was Carlos. I knew then that I had made too many assumptions about my students. They appeared as if they would be content to continue to ask "Is it this?"–type questions until they identified the mystery number.

I learned that day that playing a game also required teaching. I needed to find ways to help my students recognize good questions and think more about winning strategies when they played a game. Sports analogies seemed to help. We talked about what made a good play in soccer and how in a game you had to decide whether it was best to dribble, pass, or kick for a goal. They all agreed that they would like to be the ones to kick the goal, but sometimes that move didn't make the most sense.

Over time the students learned that I expected them to think about their moves in a game. They got better at listening to each other, rather than just thinking about the next move they wanted to make. It was exciting to see the shift; they were applying their energy and social interest to their learning.

We believe that such a change in student conduct depends on teachers reflecting on their teaching and holding high expectations for all students. We also think students need explicit access to what the expectations are and, ideally, need to share in identifying them. Further, students should have a variety of opportunities to learn ways to demonstrate positive behavior related to games and puzzles.

❯ Protocols

Working with students to create game and puzzle protocols, rules, or norms they should all follow when playing or solving helps establish and clarify expectations. You may want to display such protocols in the classroom, where they are visible to all and can be referenced during and after gaming and puzzling. Brainstorming ideas is a good way to begin.

In one first-grade classroom, a teacher gave her students the sentence stem *When playing math games and solving math puzzles, we should* . . . and recorded their thoughts on chart paper. The students' initial ideas included the following:

- Not get angry when we do the wrong thing
- Not laugh at people when they get it wrong
- Help everyone
- Not let your partners give up
- Not tell them the answer

The teacher noticed their responses emphasized what they should not do. To stimulate different thinking, she said, "What can you tell me about what you *should* do?" Their new ideas included the following:

- Do the best that you can
- Listen to everyone's ideas
- Have fun and learn
- Pay attention
- If your partners make a mistake, help them get it right
- If your partner wins and you lose, try to be nice

The teacher then asked, "What would I say you should do?" Charlie responded, "That's easy. Everyone is a winner because we are all getting better at math." In the following reflection, this teacher shares her thoughts about this discussion.

> **TEACHER REFLECTION**
>
> *My students began with so many ideas related to what they shouldn't do. I try to focus on talking about what we should do, so I decided to direct them toward that way of thinking. I was pleased with how many ideas they generated. Their*

thinking demonstrated that they knew how to apply our general classroom expectations to playing games and solving puzzles. I was particularly happy to see how many of their ideas related to working with others.

It was so good that someone suggested have fun and learn. *I want my students to associate those two behaviors—to think of our classroom as a place where they experience learning as an enjoyable activity. Charlie really sounded like me, both in what he said and how he said it. We all laughed, but I was glad that they were reminded that learning was winning.*

On another day we talked about how to help someone who seemed to be stuck. They identified three questions they could ask: Do you want a hint? Can I help you? *and* How can I help you not to give up? *I am so proud of how thoughtful these students have become. They have partnered with me and with each other as they've engaged in their own learning.*

) Game and Puzzle Manners

Your students are probably familiar with simple manners; for instance, they likely know when to say please and thank you. They also have ideas as to how to treat one another respectfully in the classroom community. But along with a game and puzzle protocol, it can be helpful to discuss specific situations that arise in relation to playing and solving, particularly when doing so in teams.

In response to teachers asking us about ways to help students apply expectations for "classroom citizens" to game and puzzle situations, we created a hypothetical "game-and-puzzle manners expert" who would give advice about how to deal with difficult situations that might arise. We asked students to take on the role of the manners expert, and we developed various scenarios for them to consider and respond to as if they were the expert. We found that with some adaptations, these scenarios worked across a variety of grade levels. Younger students often wanted to talk more about their own similar stories in relation to a situation, but once asked to think about themselves in the role of a manners expert, they were able to think more objectively about what might be a helpful reply.

One second-grade teacher invited her students to the rug area. She told them that she had notes from some other students who were concerned about things that had happened to them when they played a math game or solved a math puzzle with their classmates. She read aloud the following note, which can be found among the Manners Expert Cards in Figure 2.3 and in the Appendix on page A-3.

Dear Manners Expert,

My partner solved the math problems before I had a chance. He said he was just trying to help me, but I think I need to do the work, too, so I can learn. What should I do?

Dear Manners Expert, Sometimes my math partner and I get too silly when we play math games. I'm not sure we should be partners. How could I choose another partner without hurting my friend's feelings?	Dear Manners Expert, My partner solved the math problems before I had a chance. He said he was just trying to help me, but I think I need to do the work, too, so I can learn. What should I do?
Dear Manners Expert, The other team got really mad today when they were losing the game. Sometimes we felt like we should try to lose so they would feel better. What can we say to help them not get so mad?	Dear Manners Expert, I am very shy and get nervous when we have to work in partners. I don't even know how to ask someone to be my partner. How can you help me?
Dear Manners Expert, Sometimes my math partner gives up when we get stuck. We tried to solve a puzzle today and it was tricky. She just stopped trying to solve it. How can I help her not give up?	Dear Manners Expert, My teammates don't always clean up. They throw things in the box and leave stuff on the floor. I am tired of being the one to put everything away the right way. What should I do?
Dear Manners Expert, Sometimes I really need to solve a puzzle alone. I get too distracted in the group. I think my teacher might let me, but I'm worried about what everyone would think. What should I do?	Dear Manners Expert, Yesterday my partner had some good ideas, but our turns took forever because he wouldn't make a choice. How can we get the other team to be more patient and my partner to be faster?
Dear Manners Expert, I'm afraid to get the answers wrong and so I just say I don't know. I know my partners get frustrated and think I don't know anything. What should I do?	Dear Manners Expert, I get so excited when I win that I clap and make a lot of noise. My math partner told me I was bragging and made the other team feel bad. How can I celebrate without bragging?
Dear Manners Expert, My math partner got a wrong answer today. When I told her, she got upset with me. I didn't mean to hurt her feelings. What should I do next time?	Dear Manners Expert, My partner and I had a lot of questions about the puzzle today, but everyone we could ask looked too busy. So we just filled in numbers. What do you think we should have done?

Figure 2.3 Manners Expert Cards

Mimi said, "That sounds just like my sister. She always wants to take over everything."

Noah said, "That has happened to me. In gym some people hog the ball and don't give everyone a chance."

The teacher said, "I would like you to imagine that you are a game manners expert and that this note has been written to you. How might you respond? Talk to your neighbor about what you might say or write."

It was quiet at first, but as time passed the conversations became more and more animated. Some of the pairs were still talking about similar incidents that had happened to them, but many of the students began to talk about what to do when this type of scenario happens. When the teacher called the students' attention back to the larger group, she asked them to talk about their ideas. The teacher was impressed with their strong feelings. They all seemed to believe that the behavior needed to change. Their initial ideas were to have a rule that everyone got a turn, to tell the partner that it wasn't fair, to say, "Let me try," or to tell the teacher. Lucas added, "Last year my first-grade teacher read a book about sharing. Maybe something like that would help."

Over time, the class discussed the scenarios on several of the other cards. The teacher had the students brainstorm ideas and she listed them for all students to see. Often she had students dramatize the situation described as well as some of the solutions proposed. The dramatizations gave students the opportunity to practice their suggestions and to become more familiar with speaking up for another player or for themselves. This teacher will always remember the day she heard Emily say, "I think we may need to get the manners expert over here" while she was playing a game.

⟩ Rules Students Can Decide for Themselves

You may have had the experience of playing a particular game with new friends or with relatives who live in a different part of the country and discovering that you play the game by slightly different rules. A game of checkers may begin with players establishing whether or not you have to jump—if you can jump. It matters that the players agree, but either way, the game is still checkers. One teacher, who has shared this example with his students, replies, "Oh, that's a do-I-have-to-jump question that you decide for yourselves," when players ask questions such as the following:

⟩ How do we decide who goes first?
⟩ Does a team that makes an error lose a turn or just get corrected?
⟩ Does a roll of the dice count if it rolls off the rug?
⟩ Should both teams have the same number of turns, or does the game stop as soon as a team wins?

You'll find that most of the directions in this book that involve taking turns assume that players will decide who will play or take a particular role first. When applicable, they indicate when it is a disadvantage to go first. We found that students usually relied on rolling a die if one was already needed for the game or on rock–paper–scissors when a die was not needed for play. When we asked students how they might decide, these were the two techniques they usually shared first. Students also suggested using comparisons about themselves, such as closest birthday, oldest or youngest, name with the most letters, or first names in alphabetical order. You may want to have a conversation with your students and have them make a list of ways they could decide who will go first.

❭ Responsibility: Let Them Be the Mathematicians

As teachers, we all feel responsible for helping our students learn, but when we don't share that responsibility with them, our own behaviors could interfere unknowingly. Sometimes we might get excited when we notice a particularly good possible move in a game and point it out. Occasionally, we might draw solvers' attention to how two clues could be combined to yield important information about a puzzle. During discussions, we might answer questions to which other students could respond. All of these behaviors have the unintended consequences of limiting both our students' sense of responsibility for their own learning and their development of the habits of mathematicians. We might want to ask ourselves:

- ❭ Do I do too much telling?
- ❭ Do I believe struggle can be productive and let my students struggle long enough?
- ❭ Do I have shared goals and expectations with my students?

No matter how long we have been teaching, we all need to ask ourselves these questions. As mentioned earlier, we encouraged students to be critics and co-creators during this writing. It wasn't long before we asked ourselves why we hadn't previously engaged students in such tasks. As we realized how much mathematical thinking we were doing as we created or adapted games, puzzles, and exit questions, we were even more amazed by this omission. Being in the role of game, puzzle, or exit question creator provides students with opportunities to analyze key ideas as well as build on and critically analyze the thinking of others, which are important mathematical habits of mind. So we learned, once again, that we must make sure we are offering students every opportunity to take responsibility for their own learning. We continue to refer to this idea in each of the following sections of the chapter.

Assessing Learning and Setting Goals

Worthwhile assessment requires teachers to have a clear understanding of what is to be learned, the developmental progression in which the learning is likely to occur, and the evidence that will suggest such learning has been accomplished. It is upon such a foundation that we make decisions as to how particular games and puzzles can accomplish established goals. Creating such goals and making them clear to all stakeholders is essential.

❭ Setting Goals

Teachers, coaches, and administrators spend a great deal of time reviewing data and establishing learning goals. Such goal setting might be at a district, school, classroom, or student level. Shared ownership of such goals is necessary for their success and, yet, students are not always included. Rich Newman (2012) contrasts his son's ability to articulate goals related to his favorite video game with students' understanding of their learning goals in school. His son could identify goals for his playing and what he must accomplish to meet those goals. He could name recent improvements he had made and how they came about, as well as the next steps he would take to continue growth. He knew his current level of achievement, what he was best at, and specific aspects of the game he found challenging. How many of our students could give such detailed insight into their learning at school?

Some classrooms do have established practices for creating goals and involve students in that process. Consider the following example from a kindergarten classroom. Within the structure for setting goals in this classroom, students understood that the games and puzzles they were assigned to complete or chose to play would be aligned to their learning goals.

The teacher asked the students to close their eyes. When they did so, she chose the five-frames for the numbers one and four, shown in Figure 2.4, and covered them with a piece of opaque paper. When the students opened their eyes, she told them that she would uncover both frames for just a short amount of time. She wanted the students to look carefully at them and then think about the total number of dots they saw on the two frames.

The teacher re-covered the frames and asked, "How many dots did you see?" Deidre said she saw one and three. Abdul said he saw one and four. Paulina and Daniel each said they saw five dots. The teacher uncovered the frames again, and they all agreed that that there were one dot and four dots.

The teacher asked again about the total number of dots. Deidre made two fists and then extended one finger on one hand and four fingers on the other. She tapped her fingers as she counted from one to five and then announced that she saw five. Abdul asked if he could move one of the frames. When the teacher nodded, he turned the frame showing one so that the dot was placed directly beside the empty square in the frame that showed four. He said, "If we put this dot here, it would be full. So there are five." The teacher asked Paulina and Daniel how they knew there were five, and Daniel explained that he counted one more than four, and Paulina said she just knew.

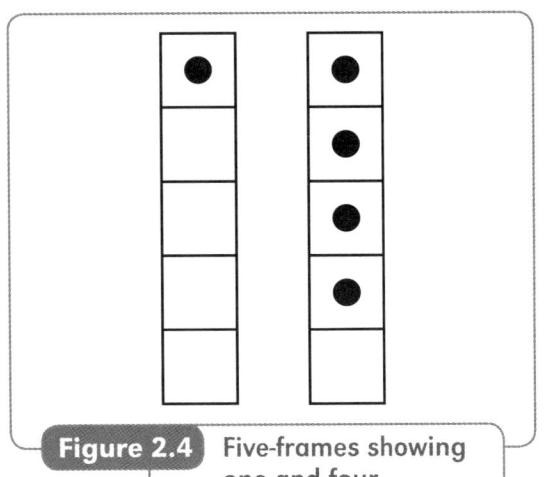

Figure 2.4 Five-frames showing one and four

The teacher repeated the process with the frames showing three and two. This time only Paulina said she knew that three plus two was five, and Daniel said he used Abdul's strategy of filling in the empty spaces. Both Deidre and Abdul represented the numbers with their fingers and counted from one to five. The teacher presented one more example with frames showing two and one. All of the students responded correctly and quickly.

The teacher asked Abdul to stay a moment while the others returned to their seats. She asked him, "What do you think you did well?" and "What do you need to work on further?" Abdul reported that thinking about filling in the empty spaces was a good idea. He also noted that he should just know what one more would be. Together they decided that Abdul would choose five-frame games that week during math choice time. He would think about knowing one more and use his thinking about filling empty spaces. The teacher recorded these decisions in her log, and she made Abdul a five-frame with four red dots and one yellow dot. She suggested that if he put this in his game folder, it would remind him of his goals. Abdul understood that these goals provided an important focus for his learning and that he would be held accountable for meeting them.

The teacher then asked Abdul to go back to his seat and she invited one of the other three students to have a conversation with her.

) Examining the Evidence of Learning

Once learning goals are established, we collect evidence as to whether or not students are meeting them. There are a variety of assessment artifacts connected to games and puzzles, but they are not always observed, recorded, or analyzed, and we all know that teachers can't be everywhere or consider every piece of student work. We believe there are ways, though, to make such evidence more visible and thus be more readily available to inform our instructional decision making. The process begins with the initial introduction of the game or puzzle.

Observations

As soon as students begin to solve or play, we encourage you to take your clipboard or tablet and become an active observer. In your first observation, you want to make sure everyone understands expectations, and you want to look for trends or patterns among the students. We suggest specific observations within the "What to Look For" section for each game or puzzle, but, in general, there are four major goals for this initial observation:

) Make sure students are following directions correctly, and intercede as necessary, before misinterpretations are practiced and become more difficult to change.
) Note what students are talking about and how they are relying on one another to reach success. What vocabulary do they use? What questions do they ask one another?
) Look for examples you want to share during the large-group debriefing regarding partial understandings, strategies, or interactions. Gain students' willingness to share, and note the order in which you want them to do so.
) Look for any challenging situations in which the students are having difficulty getting along while playing a game or solving a puzzle. Sometimes you may wish to address such a situation immediately, while at other times you may prefer to wait. If waiting seems best, you may wish to create a related manner expert card to discuss at a later time.

As opportunities to play and solve continue, perhaps with variations included, you can focus on one or two groups per session to observe more carefully, taking notes regarding evidence of learning and examples of challenges for individual students.

Student Work

We suggest utilizing two sources of work: recording sheets and oral and written responses to exit questions. Teachers found they could skim written pieces, and jot quick notes about or record oral responses, looking for correctness or patterns among the group. Sometimes they focused on particular students' learning, so their responses received more attention. For example, Figure 2.5 shows two responses to the second exit question suggested for the game *Four of a Kind* (page 116).

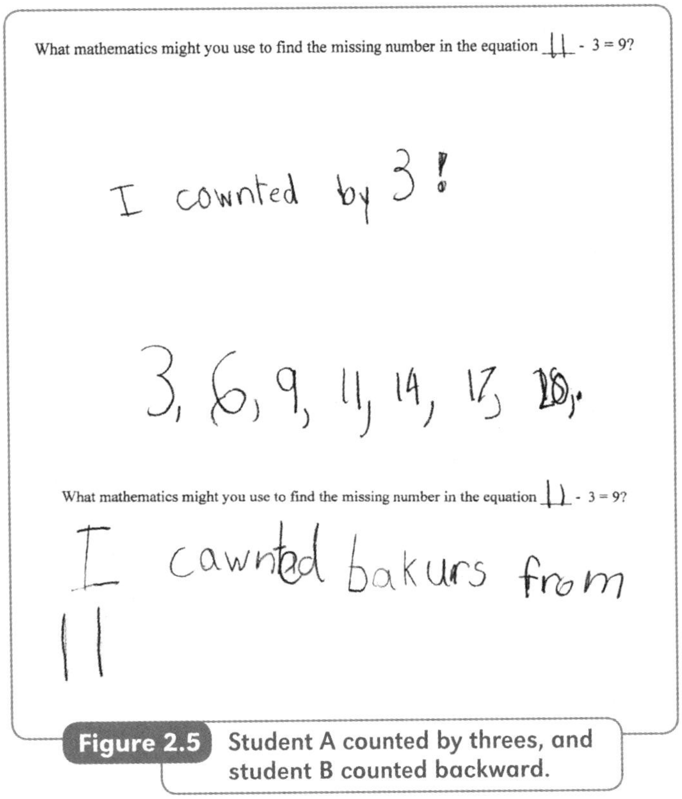

Most first graders are still learning how to explain their thinking and have many partial understandings. Both of these students identified the incorrect missing number in the equation. What feedback would you give, or what questions might you ask? Would you be more likely to ask student A to model the problem with counters or to ask if he or she could spot an error? Might you ask student B how he or she knew to start counting at eleven or present another example, such as ___ $- 4 = 7$, and ask the student to talk aloud as he or she solved it? Simply talking with either student would give the teacher and the student another opportunity to engage in a mathematical conversation.

At the end of the day, a teacher may have a pile of recording sheets and wonder what to do with them. Students often do not receive feedback about such work and often regard recording sheets as just a place to "do their math" while they are playing the game. One teacher told us, "I like to respond to the recording sheets occasionally, so students know these are more than scrap paper and that they are expected to always show their best thinking." We agree. When teachers provide feedback, students value the recording sheet more; they view it as a tool for communicating their thinking and demonstrating what they know.

There are a variety of ways you can use the exit card questions suggested in the game chapters:

Figure 2.5 Student A counted by threes, and student B counted backward.

- Pose one or two questions for students to consider in a think–pair–share format.
- Choose one question to give to the whole class, or vary the questions for different students based on their readiness or choices.
- Include one of the questions on an at-home assignment or formal assessment, letting students and parents or guardians know that games and puzzles are an integral part of their learning and require accountability.
- Offer one question twice, once after the first exposure to the game or puzzle and then again after further explorations, to document any changes. You may want to sit with students as they compare their two responses and reflect upon how much they have learned.

In response to both exit questions and recording sheets, you could do one of the following:

- Choose examples to share with students during the next math lesson.
- Sort responses quickly into three piles—not proficient; working toward proficiency; and proficient—and use the results to differentiate instruction.
- Share a simple, generic rubric (see, for example, Figure 2.6) to help clarify expectations, after which you or your students could apply it occasionally to their responses.
- Create a task-specific rubric with a colleague at your grade level and work together to apply it to students' responses.
- After reviewing them with students, place two responses to the same question or task, completed over time, in the students' portfolios to share with parents or guardians.

Name:			Date:
	Novice	**Developing**	**Strong**
Concepts	There is no evidence of conceptual thinking, or thinking is incorrect.	Partial understanding of concepts is demonstrated.	Full understanding of concepts is demonstrated.
Skills	There are several computational errors.	Most computation is correct.	All computation is correct.
Communication	There is no use, or mostly incorrect application, of mathematical terms and symbols.	Mathematical symbols are used correctly and some relevant vocabulary is included and used correctly.	All relevant mathematical symbols and vocabulary are included and used correctly.
Communication	No, or only incorrect, examples or explanations are provided.	An incomplete explanation is provided, or more clarity is needed.	Explanation is clear, correct, and includes examples.

Figure 2.6 Generic rubric for exit questions and recording sheets

Fostering Productive Discussions

Math talk is now a common term, illustrating the expectation for communication in today's math classrooms. *Accountable talk* describes the specific kind of discourse in which we want our students to participate. Such talk requires students to ask each other about their thinking, listen to each other carefully, build on the ideas of others, and use evidence to justify their ideas. It also means creating classrooms where students are willing to share their initial thinking, not just their finished products. We want to build classroom communities where errors are viewed as part of the learning process, because using errors for instructional purposes has the potential to increase students' understanding (Bray 2013). Such instructional opportunities are dependent on finding tasks that allow us to tap into potential misunderstandings and creating classrooms where errors are viewed as part of the learning process.

To some extent, such conversations, along with the teacher talk moves suggested by Chapin, O'Connor, and Anderson (2009) and Kazemi and Hintz (2014), are incorporated into many classrooms. Our task here is to think about connecting this talk to games and puzzles.

First and foremost, such conversations depend on significant tasks. Engaging games and puzzles provide something for students to talk about. Whole-group debriefing sessions offer opportunities for students to describe the mathematical understanding they gained from the game or puzzle, discuss strategies, and explain solutions. We want students to hold these same types of conversations while they are playing and solving in small groups. Sentence frames can be used to help students participate. Examples are shown in Figure 2.7.

Ask questions when you don't understand.	Can you explain why this move …? What do you mean when you say that …? Can you help me to …?
Ask about the thinking of others.	What do you think we should …? How did you decide …? Do you want to say anything about …?
Make predictions.	I predict that … If we do …, then I think …
Build on the thinking of others.	I agree with …, but … I use this idea, too, when I play, but I … _____'s idea makes me think …
Give and ask for evidence of thinking.	An example from the game we played is … I think … because … Can you tell me why it is true that …?
Look for patterns and generalizations.	Now I am wondering … This puzzle reminds me of … We could also use this when playing …

Figure 2.7 Sentence frames for games and puzzles

Meeting Individual Differences

Your assessment data will help you identify mathematical readiness levels. Several specific ways to meet such differences are suggested in the game chapters within the "Variations" sections. Ideas for meeting other differences are provided in the "Tips from the Classroom" sections. Here we would like to provide a general list of ideas.

- Create sound barriers for rolling dice by having students use rug samples, sometimes available for free at carpet stores, or appropriately sized boxes lined with felt.
- Cards and dice come in a variety of sizes, colors, and textures. Offering a variety allows students to choose what's best for them.
- Create an area to play and solve that is somewhat private, perhaps behind a bookcase, as an option for students who will have better success when working in an area that is less distracting.
- It can be challenging for many young students to sort and hold cards in their hands. You could have students place folders upright in front of their cards to keep them private. In many cases we found that players merely kept their cards in plain sight in front of them and it was fine.
- Some students may not have time to complete a written response to an exit question during class. Sometimes you may wish to collect whatever those students have completed. Other times, you or they could decide that dictating, completing, or refining responses will be their homework.
- As students play and solve together, some may prefer to respond to exit questions together as well. You can vary expectations for cooperative and individual responses. You could also allow some students to help create answers for recording sheets and responses to exit questions while relying on their partners for the actual recording.
- Students react in different ways to competitive play. Though it can be motivating for many, some become overly competitive and others resist competition or become quite anxious. Nearly all games can be played cooperatively, with the goal of trying to improve by working together.

Organizing Students for Success

After many years of thinking that the best way to introduce a new game or puzzle format was to demonstrate it to a few students and then let them introduce it to others, we now believe that a new game or puzzle format should be introduced by the teacher to the whole class or a large instructional group. You'll find that each of the games and puzzles in this book is introduced in this manner. This does not mean that you always have to play an entire game or completely solve a puzzle in a large-group setting. Sometimes you can consider a miniature version of the game or puzzle, just take a few turns or look at a few puzzle clues as a class, or explore or review related mathematics and relevant terms. We believe such introductory experiences help students in the following ways:

- They support the notion that students are part of a diverse learning community.
- They model and reinforce expectations for how best to play games or solve puzzles with team members, be patient with opponents, persevere, and be gracious winners and losers.
- They model and reinforce ways to participate in accountable talk related to games and puzzles.
- They increase the likelihood that students will be able to follow directions and meet expectations during opportunities to play or solve.
- They provide you with opportunities to gain the formative assessment data you need to differentiate follow-up opportunities for students to play the game or solve similar puzzles.

▸ Grouping Students

When games and puzzles are just for practice, teachers often group students homogeneously and have one student compete against another or have students complete a puzzle alone. When the goal is to build a deeper understanding of mathematics and mathematical habits of mind or practices, heterogeneous grouping makes more sense. As Wedekind states, "The idea of grouping kids by ability level is counterproductive to the idea that mathematicians learn from each other" (2011, 30). Playing games and doing puzzles can provide wonderful opportunities for students with different abilities to discuss a common goal. Students may be surprised to find that classmates who they thought were struggling with a concept have interesting strategies to offer or connect clues that others hadn't thought to try. Thinking of all students in the room as mathematicians, rather than just those students who seem to be the fastest and most verbal about what they know, brings new light to learning opportunities.

As with all classroom activities, you always have the choice to assign students to teams, pick opposing teams, and select locations for play, or you may choose to have the students make such decisions independently. Though we support having students make such decisions whenever possible, we often lean toward having the teacher make these decisions when the class is first considering a new game or puzzle format, assigning students in ways that will support their learning. You can display the information for all to see so that little time is lost in finding co-players and locations. If the schedule allows, we recommend that students play or solve, in teams, immediately after they have experienced a whole-class introduction.

⟩ Having Students Play and Solve in Teams

When the goal of playing a game or solving a puzzle is to deepen learning, rather than only to practice skills, we recommend having students play and solve in teams rather than as individual players. While we recognize that for a few students this may be too challenging, we consider it optimal for nearly all learners. Sometimes, particularly with puzzles, students need a few minutes to work independently before teaming, and this is fine. We've found that some students who are readers prefer to read independently and try to interpret a clue or two by themselves but then, without prompting, begin to work with their partners. Some teachers build in independent think time before cooperative puzzling, and some groups build this in for themselves. We remember Mazzie, who suggested to her group, "Let's read first and then we can talk."

Some teachers organize teams before introducing a game or puzzle; others draft team combinations, perhaps by using magnetized name cards on a cookie sheet, that can be quickly altered based on assessment data gathered during the introductory experience. Other teachers prefer to quickly organize teams with oral directions, and some allow students to organize themselves. Regardless of how teams were organized, we have been privy to wonderfully unique and informative mathematical conversations by listening to students play and solve in teams. Consider the following reflection from one of the teachers who worked with us.

TEACHER REFLECTION

I used to have students play games one against another. They didn't talk to each other very much during the games and when they did, it was usually related to who was losing or winning, but not about the mathematics involved in doing so. I was dubious when my coach suggested that they play in partner teams, one pair against another. I agreed to try it but wondered how my young students would negotiate this new arrangement. The impact was amazing. The students began to talk about mathematical ideas and, over time, discuss game-winning strategies. Our follow-up discussions in the larger group were more robust, too, as the students had already talked with their partners about mathematical ideas related to the game. So many of my students benefited from their partners' think-alouds; sometimes I felt as if I had a classroom of students and co-teachers.

⟩ Regaining Students' Attention

The engagement of a good game or puzzle can lead to a higher level of noise in the classroom than some teachers would prefer. No doubt you have one or more signals that remind students to use their inside voices or alert them to immediately stop and pay attention to you. Many teachers find that ringing a chime or striking a triangle works well. When you introduce such a system, likely during the first week of school, it is important that students practice the expected response while playing a game or engaging in solving a puzzle. Interrupting such activities may be challenging for them, especially if they feel as if they are about to win or reach a collaborative goal or solution. Assuring students that they can return to the game or puzzle after you have finished talking or at another time will often facilitate this process.

) Making Directions Accessible

Frequent requests to repeat directions can be frustrating. A reproducible with directions is provided for each game or puzzle in this book, but depending on the emerging readers in your classroom, it may or may not be helpful to students. (Regardless of the level you teach, the directions will be a valuable resource for you and can be sent home to families.) Here are some ways to provide students with access to specific game and puzzle directions:

-) Videotape yourself reading the directions and make the video available on the classroom website or available for viewing at a designated area in the classroom.
-) With permission, create a video of students playing the game that other students may watch as a way to see the directions in action. This can be particularly helpful for students who learn best by example, rather than through oral or written directions.
-) Introduce a game or puzzle a second time to a small group of students who will benefit from such a review.
-) Encourage students to draw or write a short, personalized version of the directions they can keep with them while they are playing the game for the first time.
-) Establish an area in the room where directions are available for readers and consider reading ability as an important factor in assigning groups in initial stages of playing or solving.

Organizing Materials for Success

A designated area of well-organized materials makes it more likely that students will respect and return game and puzzle components appropriately. Sending a classroom of students to play or solve in groups requires multiple copies of materials and containers. This does not mean that schools or teachers need to spend a great deal of money. Attics, basements, garage sales, families, and websites like the Freecycle Network (http://www.freecycle.org/) have much to offer, such as baskets, cookie tins, egg cartons, muffin tins, oatmeal boxes, plastic containers, shoeboxes, soap cases, and silverware organizers. What you use depends, of course, on the space and materials available to you; boxes, zip-top bags, pocket charts, and file folders are among some of the other possibilities.

If your materials are well cared for and organized, it sends a message that playing games is important and that you trust students enough to let them use materials that are attractive and well-maintained. Here are some questions and responses that can guide your thinking about the organization of game materials in your classroom.

- *Does the organization and labeling encourage student independence?* Clear picture and word labels on containers that are organized in a systematic way allow students to find and return materials more easily. Game directions along with a list of materials pasted inside each game container help ensure that components won't get lost and help students meet expectations for play and for cleaning up.
- *Does the organization support long-term game use?* Placing number cards in a soap dish within a box protects the cards, as does laminating any paper materials.
- *Does the organization support large-group use?* Large-group instruction requires multiple copies of materials. Keep multiple sets together in easily carried containers when preparing to introduce a new game. Some of these copies can then be moved to the game library for students to sign out to take home.
- *Does the organization maximize instructional time?* Having extra sets of frequently needed materials, such as dice or cards, allows students to replace missing components immediately, when needed. Same-size containers stack easily and allow games to be put away more quickly.

Working with Families

We know that the amount of parent involvement correlates to student achievement; this also applies to games and puzzles (Kliman 2006). Involving families in exploring games and puzzles related to mathematics gives both students and family members the opportunity to share the math together in a way that is different than completing a worksheet or doing problems from a textbook. Playing a game with a sibling, older or younger, allows the student to take a leadership role in explaining the rules and sharing what he or she has learned about the game or puzzle and the mathematics involved. The conversation about the math the students are learning is embedded in a meaningful context, and parents are likely to gain more information than they get in response to the question *What did you do in school today?*

One way to let parents understand the importance of playing the games or doing the puzzles with their children is to offer a games-and-puzzles party, with simple refreshments provided. You could invite family members to an afternoon or evening event in which you explain the directions to a game just as you would to students. Then parents and their children can sit together to play the game. Alternatively, the students could explain the game to their families in small groups. At such events, parents have been known to share their understanding of the math involved in a game and talk about how it is different from the way they learned it. The opportunity to address such issues and help parents understand current practices is essential, as parents identify this difference as a barrier to their involvement (Brock and Edmunds 2011). You could also invite families to bring in a favorite math-related game from their home or country of origin to share with another group.

Some families may be surprised that their children are playing games in class. You can communicate why the use of games and puzzles is important in a classroom through an e-mail message or a letter, perhaps including the directions for a popular game or the exit questions that students are expected to consider after completing a puzzle. A partial sample of such a note is provided in Figure 2.8. The more family members understand about the value of games and puzzles, the more willing they will be to spend time exploring them at home.

You can establish a lending library in your classroom, allowing students to bring home games and take responsibility for doing so. One method for providing structure to this responsibility is to include a materials list with each game packet that must be checked off when the game is returned. Students can check out the game from the classroom games library just as they would a library book, writing their name on an index card and placing it in a file or container designated for this purpose. Include a comment card in the bag so that students and their families can write something about an interesting conversation they had, something new they learned, or questions they had while they were playing. This card can be shared with the next families who explore the game.

You can also provide a game to play or a puzzle to solve as part of their homework. Require students to share something about the playing or puzzling during the next day's morning meeting. Encouraging communication about what transpired at home will provide students with an opportunity to solidify their thinking about the mathematics content. You can also ask parents or caregivers to write a short note about what they observed in their child's thinking.

Dear parent(s) or guardian(s),

Your child will work hard this year to understand the mathematics he or she is learning, compute efficiently, and be a strong problem solver. Many times we will use games and puzzles to support this learning. Students will play and solve in groups so they can talk about their thinking and discuss their strategies. I use exit cards (short questions students answer after playing a game or solving a puzzle) and recording sheets to make sure the students are recording what they are learning while they are playing or solving. Your child may bring home a puzzle or game to share with you, and I hope you will enjoy exploring it with him or her. As a result of these activities, your child will recall basic facts quickly, gain number sense, and become a stronger mathematician.

To give you a better idea of the types of games and puzzles we will be using, I am sending along …

Figure 2.8 Sample note to parents and guardians

Conclusion

We know you recognize that the instructional practices addressed in this chapter relate to all aspects of your teaching, but we hope that you can now envision them more distinctly in game- and puzzle-related situations. We also hope that you'll discover new joy in teaching mathematics when using the games and puzzles in this book or some of your old favorites with these ideas in mind. Your students will appreciate the opportunities to learn in this manner and you will see growth in their mathematical understanding and their computational fluency.

CHAPTER 3

Counting and Ordering

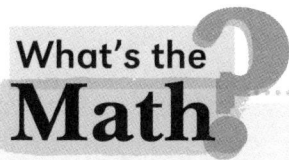

What's the Math?

Young children enter kindergarten knowing much about counting. They may be able to say numbers in order or correctly count a small set of items. Even students who have not yet learned the order of number names may have a beginning understanding of one-to-one correspondence and organizing items for comparison. For nearly all young students, though, there is still much to learn about counting. Chris Cain and Valerie Faulkner (2011) recommend that we make connections between the teaching and learning of reading and that of mathematics. Specifically, they suggest we remember that just as students can read words they don't understand, they can identify number symbols without being able to understand what those symbols represent. We should keep this in mind as well as the idea that students who are unable to read written words may already know quite a bit about stories and how books work, just as students who know few number symbols may actually know a lot about mathematics.

As counting is automatic for adults, it can be challenging for teachers to identify the complex thinking involved in the counting process. Learning to identify the value of a set of objects can involve a variety of concepts and skills, including knowing:

- the stable order of the numbers we say when we count a set;
- that each object is matched with exactly one and only one number name (one-to-one correspondence); and
- the value of a small set of objects, or a set shown in a common configuration such as on dice or dominoes, without counting or demonstrating subitizing/set recognition.

Young learners must also make connections among the various representations of numbers and be able to compare and order numbers. Using text, symbols, and graphics to represent ideas is key to mathematical thinking and communication (Diezmann and McCosker 2011). Such representations can also help students better understand numbers and the relationships among them.

One important relationship among numbers is their order, which supports students' ability to compare and order collections of objects. As order is first established through one-to-one correspondence or counting by ones, we include order games in this chapter. You may wish to have students play such games again once they can apply place-value ideas as well. Another important relationship is recognizing numbers one more than or one less than given numbers and their connection to the order of the count. A firmly established order of the count also facilitates students' ability to count on from a number other than one.

Count 20

Why This Game or Puzzle?

The development of an internal linear representation of numbers is essential to success in mathematics (Booth and Siegler 2006), and playing games with numbered game boards can help develop such a model. These boards show numerals in order, while also helping students visualize distances between numbers. As turns in such games begin from the number reached on the previous turn, players experience counts that do not always begin with one. *Count 20* presents such a board and provides many opportunities for students to practice their counting skills.

To play, each team begins with ten counters, and rolls of the die determine how counters are moved around the board. On each turn, teams decide which of their counters to move. If a team's counter completes a move and lands on a space already occupied by another counter of any color, the team can take that other counter and place it in its cup. This rule encourages players to think about which counter to

Count 20 Game Board

1	2	3	4	5
20				6
19				7
18				8
17				9
16				10
15	14	13	12	11

move and motivates them to count to predict outcomes. If players end a turn on 10 or get to 20, they can place their counter in their cup, which indirectly reinforces the idea that tens are special. At the end of the game, the team with the greatest number of counters in its cup wins.

Math Focus
› Counting
› Predicting outcomes of counts
› Comparing collections of counters

Materials Needed
› 20 small counters, 10 in each of 2 colors, per group
› 1 die per group
› 1 cup per team
› 1 *Count 20* Game Board per group (page A-4)
› Optional: 1 *Count 20* Directions per group (page A-5)

Directions
Goal: Have the most counters in your cup at the end of the game.
› Decide which team goes first. The other team chooses the color of counters for each team.
› Each team begins with ten counters of the same color.
› On each turn:
  › Roll the die and choose a counter to move.
  › Count forward the number of spaces shown on the die. As one team member moves the counter, the other says the numbers on the spaces aloud. If there is another counter of either color on the number at which you finish, put that counter into your cup and leave your counter in that space.
  › If your move ends on 10, put your counter into your cup.
  › If your counter gets to 20, put it into your cup.
› The game ends when one of the teams does not have any counters to move.
› The team with the most counters in its cup wins.

How It Looks in the Classroom
One kindergarten teacher introduces a simpler version of this game with a game board drawn on a large piece of butcher paper. She gathers the students around the paper and has the students say the numbers they see, in order. Then she points to different numbers

and asks students to name them. She is not surprised when there is some hesitancy in their responses. She knows that identifying the numerals in order does not mean that students can recognize the numerals when presented randomly.

Next she points to the students on the right side of the room and tells the students they can go first. She then asks a student on the left side to choose the color of the counters each team will use. She gives each team a cup, four counters, and a die. She tells them that on each turn they will roll the die to find how many spaces they can move and that they can choose to move any of their counters. She further explains that their goal is to get as many counters as they can into their cups. Then she turns over a piece of chart paper and reveals the three ways to get a counter in a cup:

- End on 10 and put the counter into your cup.
- Get to 20 and put the counter into your cup.
- End a move on a space where there is another counter and put that other counter into your cup.

The students play as a large group. During the game the teacher occasionally asks, "What number would you like to roll? Why?"

Then the teacher shows them the *Count 20* Game Board (page A-4) and tells them that they are going to play again, but this time there will be only two players on each team, and they will start with ten counters each. The students are excited by the thought of starting with so many counters, and Jared declares, "I want to get them all."

Tips from the Classroom

- Some players will always move one counter along the board until it is placed in a cup. Others may appear to make random choices as to which counter to move. After students have had a few opportunities to play the game, demonstrate game situations that model what it looks like when players consider alternative choices.
- As you observe students playing the game, when appropriate, challenge players to predict where their counter will land before they count.

What to Look For

- How do players determine what the number on the die represents?
- Do students demonstrate consistent one-to-one correspondence?
- What strategies do players use to keep track of their counting? For example, do they touch each space as they count it?
- What conversations do players have about which counter to move?
- Do players count to predict which move is best or to determine the number they would like to roll?
- At the end of the game, how do students compare the number of counters in each cup?

CHAPTER 3
Counting and Ordering

Variations
- You can create a board that has fewer or more numbers on it.
- Players can begin with fewer counters.
- To give students more opportunities to make decisions, have them roll two dice on each turn and decide which one to use.
- You may not wish to let teams take the opposing team's counters. Instead, have teams put only their own counters in their cups.
- To expand players' strategic options, you could allow students to split a move between two counters.
- To continue interest in the game, you can add a new rule, such as *You need to roll a 1 or 2 to place a counter on the board*.

Exit Question Choices
- How did you decide which counter to move?
- How did you decide which team had more counters?
- Project a copy of the game board and pose one of the following questions:
  - You are on 7 and roll a 3. Where will you land? Explain your thinking.
  - There is a counter on 12, and you have a counter on 9. What number do you hope to roll? Why?

A student response to the question *You are on 7 and roll a 3. Where will you land?* is shown in Figure 3.1. The student counted on from seven aloud, while putting out a finger for each of the three numbers she named. She then wrote that she would get to 10 and recorded the numerals 7 through 10. She circled her 7 and said, "I want to show I started here."

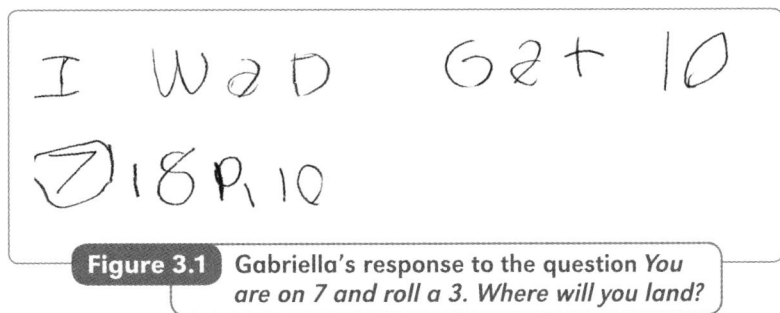

Figure 3.1 Gabriella's response to the question *You are on 7 and roll a 3. Where will you land?*

Extension
Have students make up their own rules for the *Count 20* board. When we field-tested this game, one group of students suggested *When you roll a 2, you count back two*. Another group created the rule *When you end on a space with another counter, the counters "piggyback," both become yours, and they travel the rest of the board together*. This rule worked particularly well when using linking cubes as counters and made it exciting when the tower was recaptured by a third counter.

Number Jigsaw

Why This Game or Puzzle?

Young children need many opportunities to develop and internalize multiple representations of numbers (Jung 2011). For this puzzle, students get nine individual puzzle pieces that must be matched on all sides that touch. Different representations of the same number make a match. To reach the solution, students must decide how to sort and organize the information, which are key skills required by many real-world situations.

Two versions of the puzzle are provided. The first puzzle requires students to match numerals, word names, and collections of objects. Numbers are ten or less and sets are often shown within a ten-frame or a common set configuration. The second version also includes some teen numbers as well as the phrases *1 more than* and *1 less than*. Visual images in this puzzle may also suggest thinking about "more than" and "less than" relationships.

Math Focus

- Counting
- Matching number names, symbols, and sets
- Recognizing numbers that are one more or one less

Materials Needed

- 1 *Number Jigsaw* Puzzle (A or B) per pair (page A-6 or A-7)
- Optional: 1 *Number Jigsaw* Directions per pair (page A-8)

Directions

Goal: Arrange the puzzle pieces so that the numbers shown on all touching sides match.

- Work together.
- Place the nine puzzle pieces together to make a square.
- The numbers shown on all touching sides of the puzzle pieces must match.
- Check to make sure you have matched each side correctly.

How It Looks in the Classroom

One kindergarten teacher displays four cards similar to those in the puzzle. (See Figure 3.2.) The teacher points to the first card and asks Krista to choose and point to one of the numbers shown. She points to the 6 and proudly announces, "Six, because I'm six." The teacher then asks the students to look for another puzzle piece that shows six. After waiting a bit, she asks Hanis to point to the match. Once other students agree that there is a match, she puts these pieces together, with the two representations of six touching.

The teacher then says, "Turn and talk with your neighbor about where you think the other pieces will go."

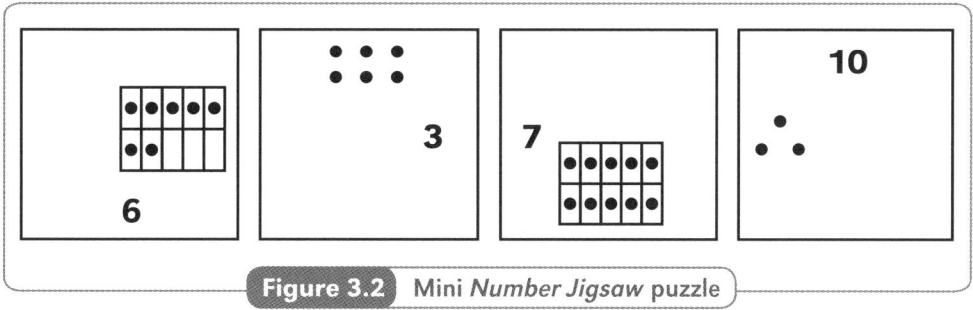

Figure 3.2 Mini *Number Jigsaw* puzzle

The students agree on the placement of the other pieces, as shown in Figure 3.3. The teacher asks the students to describe the puzzle, and they talk about matches on all sides. Lizbeth reports that the pieces make a square, and Jayden notes that there are not any numbers or pictures on some sides. The teacher tells the students that they are going to work with partners to solve a puzzle with more pieces and shows them the nine pieces arranged in a square, but not matched correctly, to give them a visual model of how the puzzle should look when completed. She says, "Remember that you need to put these pieces together so that the numbers on all sides match."

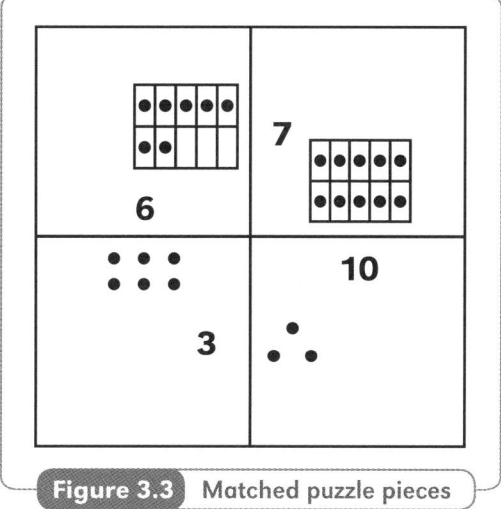

Figure 3.3 Matched puzzle pieces

Tips from the Classroom

- Group readers with nonreaders and lead choral reading of number names before students solve on their own.
- Students may have difficulty keeping the puzzle pieces in position once they find a match. Stapling the puzzle pieces onto sticky notes will keep them stable, but moveable. Placing them on a piece of felt can also be helpful. To provide more

support for those students who struggle with visual organization, you could put strips of Velcro on the backs of all the cards and also on a piece of cardstock to show the spots where they should place the cards.
- A few students may find it helpful to have counters and ten-frames available to represent what is shown and then count the concrete objects. Others may appreciate a list that they can refer to that shows numerals along with corresponding written names and sets.
- Some puzzle sides have more than one match. Students need to choose the piece that allows all sides to match. Some students did not notice this and thought there was an error in the puzzle when they were left with unmatched pieces. You may need to assure them that the puzzle works, or in some cases, ask them to see if there are pieces that could be exchanged.

What to Look For
- Do students seem to prefer one representation over another? That is, do they often begin with a ten-frame representation, a set not shown on a ten-frame, or a symbol?
- What evidence do you observe of subitizing/set recognition?
- What counting concepts or skills do players find most challenging?
- What strategies do students use to identify a number that is one more than or one less than a given number?
- Do partners work together to solve the puzzle or do they work individually to match pieces and then share what they find?

Variations
- You can place a star on an edge of each corner piece to make the solving somewhat easier.
- You can create versions of this puzzle that focus on place value or addition and subtraction.

Exit Question Choices
- You are making a number puzzle. How might you draw a set for the number 8?
- How do ten-frames make it easier to count?

Extension
Have a *Number of the Day* routine for a week. Sometime during the day, each child adds a visual representation of that day's number to a class chart. Review the images with students at the end of the day. Ask questions such as *How are these images alike? How are they different?* and *Which images make it easier to count? Why?*

Nim

Why This Game or Puzzle?

Meagan Burton (2010) recognizes the use of math games as an important instructional strategy, identifying their ability to support both fluency and logical thinking. *Nim*, a game thought to be hundreds of years old, provides young learners many opportunities to count, as well as to note patterns, make observations, and draw conclusions, all behaviors essential to strategic thinking. The counting aspect of the game is aimed at kindergartners, while the qualitative reasoning can provide a challenge for learners of any age. There are several variations to the game, allowing interest to be maintained.

In this simple version of *Nim*, fifteen counters are placed in a row. Players remove one or two counters on each turn, and whoever takes the last counter is the winner. Once students play the game several times, they will begin to recognize patterns that can inform the choices they make. With this set of rules, it is possible to always win the game by playing second and always making the opposite choice of your opponent. For example, if your opponent takes one counter, you should take two. Some students will be able to discover this strategy eventually; most will begin to recognize which team will win when only a few counters remain. Students' ability to make predictions is likely to improve if they complete the *Nim* Recording Sheet (page A-9). Though we recommend you ask questions such as *Is there a point in the game when you can predict who will win?* we cannot overemphasize the importance of allowing students to discover patterns and winning strategies rather than be told about them.

Math Focus

- Counting
- Recording numbers
- Making predictions

Materials Needed
- 15 counters per group
- 2 *Nim* Recording Sheets per team (page A-9)
- Optional: 1 *Nim* Directions per group (page A-10)

Directions
Goal: Take the last counter.
- Place fifteen counters in a row.
- Decide who goes first. This is Team 1. The other team is Team 2.
- Take turns.
- On each turn, take one or two counters.
- After each turn, both teams write the number of counters taken and the number of counters left on their recording sheet.
- The team that takes the last counter wins.
- Play again. This time, Team 2 goes first.

How It Looks in the Classroom

Demonstrate the game to the whole class by projecting fifteen counters in a row and asking a student to play the game with you. Explain that on each turn, players will take one or two counters from the row and that the winner is the player who takes the last piece. Play a game with the student volunteer. Make random moves so as not to demonstrate a winning strategy.

Next, project the recording sheet and ask, "What do we write if we start the game and the first player takes away one counter?" Have students talk in pairs and then invite about three students to share their strategies for deciding how many counters would be left. Then have students restate the rules of the game to you before having them play in teams.

Tips from the Classroom
- Have students play a round or two of the game, to become familiar with it, before completing recording sheets.
- Some students will find it helpful to have a cup or container in which to place removed counters.
- Playing a game with a set of six or seven counters may help students make predictions about who will win.

What to Look For
- How do players decide how many counters to take on their turns? What evidence do you observe that players consider the consequences of their choices?
- At what point in the game do players recognize who will win?
- How do students determine the number of counters left after one or two more counters have been removed?

Variations

- Change the total number of counters.
- Allow teams to take one, two, or three counters on each turn.
- Have players place one or two counters in a row, rather than take them, with the goal being to place the fifteenth one to be the winner.
- Have the team that takes (or places) the last counter lose rather than win.

Exit Question Choices

- There are ten counters left. You take one. How many counters are left now?
- There are four counters left. It is your turn. What should you do? Why?

We were not certain how young learners would react to the second exit question, but we quickly became impressed by their ability to discuss game scenarios. Because of the reasoning involved, it is best to have students respond orally. One teacher invited a parent to talk with the students individually and record their responses. A strong kindergartner's response is shown below. Note that this student extended the question by also talking about what would happen when three counters were left.

- If there are 4 left, if I took 1, then you took 2, then I took 1, then I win. If I took 1 and you took 1, then I took 2, then I win again.
- If it is your turn and there are 3 left, then I win too, because if you take 2, then I take 1. If you take 1, then I would take 2.

Extension

Have students play *Nim* with a family member or caregiver and share their experiences of doing so at a morning meeting.

Why This Game or Puzzle?

Terrific for playing as a whole class during transition times, this game can also be played in small groups or pairs. The teacher or student leaders choose a number. The players' task is to ask questions that have yes-or-no answers until they can identify the mystery number.

The mathematics of the game focuses on comparison of numbers, which is most often associated with a number line model.

Other questions might refer to the number of digits or whether it is odd or even. The game also provides an opportunity for thinking about strategy through conversations related to questions such as *What makes a good question? What did you just learn from this answer?* and *What's a good way to organize the information?* Such thinking is important, as the way mathematics is explored in classrooms should reflect how it is used in the world (Van de Walle, Karp, and Bay-Williams 2013).

The recording sheet is optional, as some younger students may not have the writing skills to complete it.

Math Focus
- Comparing numbers
- Identifying number names
- Using "less than" and "greater than" signs to record information

Materials Needed
- Optional: 1 *Mystery Number* Recording Sheet per group (page A-11)
- Optional: 1 *Mystery Number* Directions per group (page A-12)

Directions
Goal: Name the mystery number.
- Choose two game leaders.
- The game leaders write a number and keep it private. They tell how many digits are in the number.
- The other players take turns asking yes-or-no questions. The leaders answer.
- Players record questions and answers on the recording sheet.
- Ask questions until a player names the mystery number by asking *Is the mystery number _____?*
- The leaders show their written number when a player names it correctly. If players make an incorrect guess, they should continue asking questions until they make a correct one.

How It Looks in the Classroom

One second-grade teacher begins by asking the students, "Do you think you could solve a mystery?" Following a resounding positive response, she continues, "I am going to think of a mystery number and you are going to be detectives, trying to identify it. You can gather clues about the number by asking me questions. I can answer only by saying yes or no. When you think you have enough clues to solve the mystery, you can ask a question such as *"Is the number eighty-four?"*

The teacher then begins the game by saying, "I'm going to write the number on this piece of paper, so I can show it to you when you guess correctly. The number has three digits. Who has a yes-or-no question to ask?"

Emmy says, "Is it 175?"

The teacher shakes her head while saying, "No." She then tells the students that like good detectives, they should make notes about what they learn. To model a possible recording, the teacher writes *175* on the board and crosses it off.

Harry asks, "Is it 199?" Again, the answer is no, and the teacher records this information in a similar manner.

Joseph asks, "Is it more than 200?"

This time the teacher replies, "Yes," and then asks, "How should we record this information? I don't want to write the numbers *100* to *200* and cross them all out!" Eventually the class decides to record *>200*. The teacher is not surprised that it takes many questions to identify the number 243 and recognizes that students will gain efficiency with more opportunities to play the game.

The teacher then organizes students in groups of four. She explains that two players will ask questions and two will answer them. She distributes recording sheets for them to use. (See page A-11.) She then says, "After you play one game, play again, but change who asks and answers questions. After you play, we will talk about what makes a good question to ask."

Tips from the Classroom

> A hundreds chart can be helpful for students playing with numbers that have fewer than three digits. They can use it as a recording sheet or as a visual guide. With three-digit numbers, students can record information on an open number line. Such references can be particularly helpful for checking the correctness of leaders' responses during and after the game.

> Students can use whiteboards to record their work and then hold them up for others to see.

> Encourage students to stop at some point in the game to reflect on what they know and decide what information they want to know next.

- Some students may need question starters, for example, *Is your number less than _____?* You could post question starters on chart paper to make them available to the whole class.
- The assessment data gained from students' completion of the recording sheets is informative, but it is not necessary to complete such a recording for every game. As students become more familiar with the game, they may prefer to use their own recording methods.

What to Look For
- How do students record the information they gather?
- Are students asking questions with answers they could have deduced from previous questions and answers?
- What kinds of questions do students ask? Are their questions broad or specific enough? Which students need more support to recognize questions that are better than others to ask?

Variations
- Provide a specific range of numbers to be announced at the start of the game, for example, *The mystery number is greater than 200 and less than 500.*
- To build both a cooperative and a competitive spirit to the game, have students keep track of the number of questions asked and establish the class or group goal of trying to guess the number with fewer questions the next time students play.
- Have the leaders give two clues about the mystery number over the course of the game.

Exit Question Choices
- You know the number is greater than 30 and less than 50. Should you ask, "Is the number less than 60?" Why?
- You know the number is between 300 and 600. What question might you ask next?

Extension
Some students may wish to prepare a dramatization of a game in which one player is always asking questions such as, *Is it 53?* The other players need to suggest other questions to ask and explain why they are more useful.

Order Up

Why This Game or Puzzle?
Research suggests that we have a mental image of numbers similar to that of a number line (D'Arcangelo 2001). So although base ten knowledge may make it easier to order numbers, the order is established in a ones view. Students can recognize that numbers in the hundreds come after numbers in the tens without necessarily recognizing that there are 10 tens in 1 hundred. Students may use a variety of concepts to put numbers in order. They might think about the number of digits, the order of the count, or comparisons with benchmark numbers.

Math Focus
- Reading numerals
- Comparing and ordering numbers

Materials Needed
- 1 deck of *Order Up* Cards per group (pages A-13–A-15)
- Optional: 1 *Order Up* Directions per group (page A-16)

Order Up Cards

7	8	17	18
24	26	29	30
31	42	45	56

Directions
Goal: Put five numbers in order from least to greatest, left to right.
- Shuffle the cards and place five cards faceup, from left to right, in front of each team. Teams may not change the order of the cards in their "hands."
- Put the other cards facedown in a deck.
- Choose a player from each team to do rock-paper-scissors. The winning player's team goes *second*.
- The first team chooses the top card from the deck. The team may trade this card for one of its five cards. If the team makes a trade, it places the card from its hand faceup next to the deck in a discard pile. If the team does not make a trade, it places the card drawn from the deck faceup in the discard pile.

> - The players take turns, picking a card from either the deck or the discard pile. Then they place either a card from their hand or the card that was drawn faceup on top of the cards in the discard pile.
> - The first team to get its five numbers in order from least to greatest, left to right, is the winner.

How It Looks in the Classroom

One first-grade teacher shows three numbers—13, 82, and 78—and asks students to talk in their groups about how the numbers should be written in order, least to greatest, going left to right. She overhears one group discussing whether eighty-two or seventy-eight is less. She notes that Christopher suggests that seventy-eight is more than eighty-two, because eight is more than two. Then Jamie convinces Christopher that "all of the eighties come after the seventies." She hears Rosmailyn tell her group, "Thirteen is first because it is in the teens, and they are really small numbers."

The teacher brings their attention back to the large group and asks, "What could we do to prove the order of the numbers?" Henry suggests that they count. The teacher agrees and identifies one student to stand when the class says thirteen, another when the class reaches eighty-two, and another when the class gets to seventy-eight. The counting process proceeds, and all agree on the order of the numbers.

Then the teacher introduces *Order Up*. She deals two sets of five cards faceup under her projection device so all students can see. She puts the remaining cards on the table, facedown in a deck. She tells them that the goal is to order the cards from least to greatest, but the cards can't just be moved around. She explains that she is going to pick the top card from the deck and see if she wants to exchange it with one of her five cards. She draws a card and says, "I want to trade 17 with 82 because I think 82 is too large for the first number." Figure 3.4 shows the cards and the trade she will make. After she trades, she puts the 82 faceup to start a discard pile.

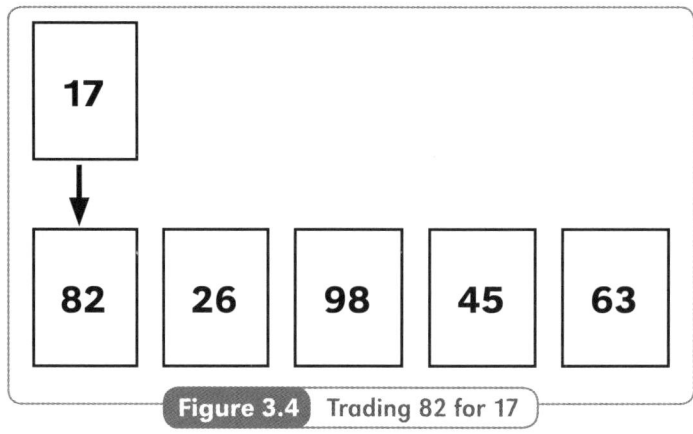

Figure 3.4 Trading 82 for 17

She then explains that the students, playing as a class, can now draw a card and see if they want to use it. She tells them they can draw the top card from either the deck or the discard pile, as she points to each one. She invites Samuel to choose for the class. She continues taking turns with the class until they finish the game.

The class plays another game. This time the teacher has students talk about their hand before they play. She asks, "Which numbers are already in order? Which numbers do you want to change first?" After playing this game, the teacher has students summarize the rules and then play in teams.

Tips from the Classroom

- Some students will benefit from access to a number line or hundreds chart.
- Encourage students to talk about the order of the cards before the game begins. Such conversations will help them note relationships among the numbers as well as develop a strategy for the game.

What to Look For

- What language do players use as they talk about the order of the numbers?
- Are students developing strategies or randomly replacing numbers?
- What, if any, misconceptions do you observe?

Variations

- Change the goal to placing four numbers in order, or increase it to six.
- Instead of stacking the extra cards in a deck, students can place the cards facedown in an array like in a concentration game. They can place discarded cards facedown in the array, and players can choose any card on their turn.
- Players can keep the extra cards in a deck but place all discarded cards faceup in view. Players may then choose a card from the deck or any of the discarded cards.

Exit Question Choices

- Should the card showing 59 go to the left or the right of the card showing 63? Explain how you know.
- You have these cards in this order: 8, 56, 13, 31, 59. You draw a 25. For which card would you trade it? Why?

Extension

Start a bulletin board with pictures that show real-world examples of numbers in order, for example, buttons in an elevator or numbers in an apartment building. Encourage families to send in pictures to add to the board.

Online Games and Apps

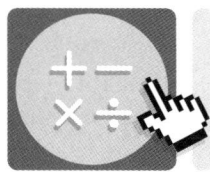

Technology can be used to support a variety of counting and ordering skills. For example, there are many games that require players to identify a particular number symbol before a bubble bursts or the time runs out. While often engaging for students because of the game theme or the competition, many of these games do not utilize the power of the computer to assist students in learning concepts related to counting.

In contrast, games that provide students with visual models of number relationships can offer learners access to conceptual understanding. Choices of difficulty levels in a game format also allow for differentiation, which is less often available when students engage in activity-based apps. Some examples include the following:

- Games that are engaging because of their context and necessity to problem solve offer variety in how they are played and provide a number of ability levels. One example is Bugs and Buttons. This fee-based app, found at http://www.littlebitstudio.com/bugsandbuttons.html, has a number of different types of games, one of which focuses on counting. Players place a finger on each bug, one at a time, using one-to-one correspondence to count as the bugs move around on the screen. As each bug is touched, it stops moving, and the number of bugs counted thus far appears on the screen. When all bugs are counted, the player chooses the number symbol that represents the total, from five symbols listed on the side of the screen, and scores 1 point if it is correct. As the player becomes more adept at number recognition, the range of numbers increases, moving from one through five to two through six, and so forth. There are a number of other games available on this app, one of which involves sorting and counting. The sorting game requires players to sort buttons into one or more boxes, and as the buttons are placed, they are counted. The backdrop of this game allows for the buttons to be counted on a grid, much like a ten-frame.

- Motion Math Zoom is a fee-based app found at http://motionmathgames.com/motion-math-zoom/ that involves players in finding the correct place for a number on a stretchable number line as it is moving. The numbers are represented as animals of various sizes, such as dinosaurs for large numbers and amoebas for small numbers. Players score points for moving along the number line until the number in the bubble is appropriately placed.

- Caterpillar Ordering, a free game from the Topmarks website, at http://www.topmarks.co.uk/ordering-and-sequencing/caterpillar-ordering, provides students with the opportunity to either order or sequence numbers. The ordering game allows players to choose from a variety of number ranges and to order them either forward or backward. The sequencing aspect of the game engages partners in simple sequences and pattern recognition. Players are shown the caterpillar with a first and last number on his body, and there are three open spaces to place numbers. There are five numbers to choose from to fill in these open spaces.

When playing online games, students should still be expected to provide evidence of their learning. For example, if playing the number line game, students can record their moves in order to practice writing numbers and to allow the teacher to assess their learning. Students may also answer questions about what they learned while playing a game, such as *What is a number that is on the number line between 6 and 10?* or *How did you decide where to place the buttons when you were sorting them?*

CHAPTER 4

Base Ten Numeration

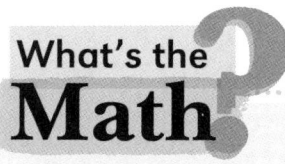

Moving from a view of numbers as ones to understanding our base ten representation of numbers is a key goal of the primary grades, one that underpins students' number sense and many computation strategies. In first grade students are expected to understand place value with numbers to 100, and in second grade, this expectation is extended to numbers to 1,000. Understanding place value includes the ability to understand regrouping and to make connections among different representations of numbers. Place-value skills include knowing the names and values of the places and identifying a digit in a particular place.

Ideas related to place value can be challenging to teach and to learn (Ross 2002). It is a challenging idea that 10 ones can form a new unit that is 1 ten, and that 1 ten can simultaneously represent 1 ten and 10 ones. A complex concept for most young students to develop, it is particularly problematic for those with learning issues (Thouless 2014).

Students need many opportunities to explore numbers in flexible ways in order to recognize the many forms in which numbers can be written and still have the same value—for example, that 3 tens and 9 ones is equivalent to 39 ones. Though manipulatives are considered necessary, but not sufficient (McNeil and Jarvin 2007), students should have ample time to explore models such as number lines, number charts, and base ten models and to gain a sense of which visual models best apply to particular situations.

The games and puzzles provided here engage students in these fundamental ideas and allow concepts related to place value to deepen. Some of the games and puzzles connect to addition and subtraction, though conceptual understanding of our base ten number system is always the main focus. Finding the number that is 1 ten more than 20, for instance, can be thought of as an addition problem or as renaming 20 to 2 tens, counting up to 3 tens, and then recognizing 3 tens as 30.

The chapter opens with *Win 1,000*, an adaptation of a classic game using base ten materials, because it offers concrete models of regrouping. While it is sometimes thought of as an addition game, here the focus is on the physical modeling of renaming numbers. The game is easily adapted to *Win 100*, appropriate for students in grade 1. As kindergarten students do not usually learn about place value, except, perhaps, to recognize the teen numbers as composed of 1 ten and some ones, this chapter is appropriate only for kindergartners working above grade level.

Win 1,000

Why This Game or Puzzle?

Constance Kamii (2014) argues that children develop number concepts from indirect instruction rather than direct teaching; that is, children need to construct number concepts instead of being told about them. Kamii also notes the benefit of games over worksheets, "as games teach arithmetic more indirectly than worksheets" (77). *Win 1,000* (or the variation, *Win 100*) provides a concrete model for trading 10 tens (or ones) for one of the next-size units, in this case, one hundred (or 1 ten).

In this game, teams roll two dice of different colors, one representing tens and the other, ones. The numbers rolled indicate the base ten blocks to place on their place-value chart. The colors represent a different place value for each team, which potentially increases students' awareness of the different magnitudes of tens and ones. Play continues with teams remembering to make trades whenever possible. The first team to have 1,000 (or more) shown on its chart wins.

The game is easily adapted to *Win 100* (see "Variations"). Frequent play of either version supports learning of an algorithm involving adding by place. Students who exhibit the ability to determine what the final value will be without manipulating the blocks are demonstrating their readiness for more abstract approaches.

Math Focus

- Understanding that the digits of a three-digit number represent hundreds, tens, and ones
- Understanding that 10 hundreds is equal to 1 thousand, 10 tens is equal to 1 hundred, and 10 ones is equal to 1 ten
- Reading and writing numbers to 1,000

Materials Needed

- 2 dice, each a different color, per group
- 1 *Win 1,000* Place-Value Chart per team (page A-17)
- A collection of base ten materials per group
- Optional: 1 *Win 1,000* Directions per group (page A-18)

Directions

Goal: Show 1,000 (or more) on your place-value chart.

- One color die represents tens, and the other die represents ones. It is different for each team. For example, if Team 1 chooses red for ones, red represents tens for Team 2.
- Teams take turns rolling the dice.
- Team 1 looks at the dice and places that number of tens and ones on its place-value chart. If it can make a trade for a greater place value, it must do so. For example, if it has 10 ones, it must trade them for 1 ten. Team 1 reads the number represented on its chart.
- After the players on Team 2 agree with the number Team 1 read, they use the same roll of the dice to place their number of tens and ones on their place-value chart, trade if they can, and read their number to Team 1.
- The first team to show 1,000 or more on its place-value chart wins.

How It Looks in the Classroom

One second-grade class is familiar with the base ten materials. The students have built a tens piece from ones pieces placed end-to-end and a hundreds piece from the tens pieces placed side by side. As a class they have placed a ones piece on their place-value charts and added one more until they've reached one hundred, trading as necessary, and they have recorded (on chart paper) the numerals the blocks represented as they did so. Then in groups of two or three, they each represented a different hundred (100 through 900) on their place-value charts and added one more until they reached the next hundred, again listing the numbers as they did so. These lists hang in the room for all to see.

The teacher begins this math lesson by reviewing the renaming process. She gives each pair one place-value chart (page A-17) and provides each group of four students with a set of base ten materials to share. The teacher asks the students to use the blocks to show 145 on their charts as she projects her chart for all to see. Once they do so, she asks them to show one more, to say the number shown on their charts, and to write the number. After they have done so, the teacher models the process. When she asks students to show one more than 149, she observes carefully. Some students regroup automatically, some hesitate briefly, and some students gently remind their tablemates that a trade can be made. The teacher shows the regrouping on her chart and then asks students to show 397 on their charts. Students continue to show one more until they are representing 401.

Next the teacher projects a variety of numbers on her chart. She asks students to predict the number that will be represented when they show one more; then they check their predictions by placing another ones piece on their charts and, if necessary, regrouping. She notes that a few students only regroup ones to tens when asked to show one more than 499, and some record 4109 as the number represented. After another two examples and a reminder to "always trade up, if you have ten," students are successful. The teacher is not surprised that this review is required. She knows that these students need many explorations to fully understand how our number system works. She reminds the students, "You can always check the list we made of the numbers 0 through 1,000 when you are not sure what number is next."

The teacher then explains the rules to the game *Win 1,000*. They play a few rounds as a class and then teams of two play games against each other.

Tips from the Classroom

> As students become familiar with the game, some players prefer to have teams place their blocks simultaneously rather than wait for each other. Be sure each team still reads its number to the other team at the end of each turn.

- We found some players chose to line up the ones blocks beside a tens block to decide if there were enough to make a trade. Remind them to make sure the endpoints of the row of ones blocks and of the tens block match.
- The ones blocks are often referred to as units, but a group of ten within any place value is traded up for the next-size unit. Therefore, we encourage you to refer to hundreds, tens, and ones.
- Encourage opponents who believe an error has been made to ask a question that might trigger the other team's thinking, instead of just telling the players that they are wrong. We found it helpful to have students brainstorm ideas for these questions before playing, such as *Are you sure your turn is over? Is there a trade you could make?* and *Did you want to recount those tens?*

What to Look For
- What strategies do students use to know when to trade?
- Do students read the numbers shown on their mats correctly?
- What evidence do you observe that indicates students understand the importance of whether a number rolled stands for tens or ones?
- When teams think an error has been made, do they just tell the players that they are incorrect or do they ask their opponents a question to help them identify it themselves?

Variations
- Have students play *Win 100*, with both dice representing ones.
- Have teams take turns rolling and placing the blocks, rather than use the same roll in different ways.
- If students require more challenge, as soon as they roll the dice, ask players to predict what will be on their board after their turn is completed.

Exit Question Choices
- You have 3 hundreds, 2 tens, and 7 ones on your mat. What is the value of the number these blocks show?
- You have 5 hundreds, 12 tens, and 13 ones on your mat. You trade. What is on your chart now?

Extension
Have students determine the least number of rolls needed to win the game.

Number Sort

Why This Game or Puzzle?

Classification involves grouping things based on their attributes. Critical to classification is the ability to identify the "rule" that determines whether something is included or not (Marzano, Pickering, and Pollock 2001). Venn diagrams are a common graphic organizer used for comparing and contrasting ideas. Two-ring Venn diagrams identify four regions, as shown in Figure 4.1.

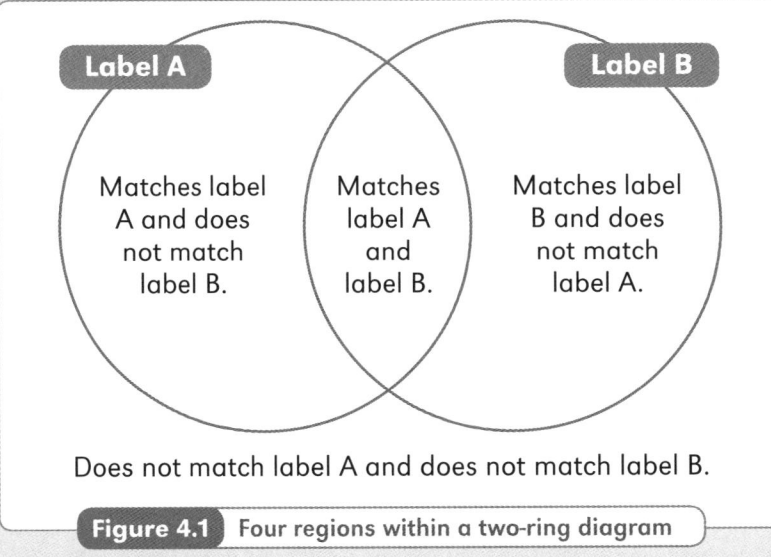

Figure 4.1 Four regions within a two-ring diagram

In this noncompetitive game, students classify the numbers one through thirty-five according to descriptions on label cards, for example, *Has a 2 in the tens place.* One team of students secretly chooses two label cards, one for each ring of a Venn diagram. The students on the other team, the guessers, choose numbers for the team that knows the labels to place correctly in the diagram. When the guessers think they can identify the labels, they should do so. They must identify both rules correctly; if they don't, the other team will simply say, "No."

Math Focus

- Classifying numbers by given attributes
- Identifying digits in the ones and tens places
- Comparing numbers
- Organizing data in a Venn diagram

Materials Needed

- 2 large strings tied in loops or 1 large Venn diagram drawn on chart paper per group
- 35 chips, labeled with the numbers 1–35, per group
- 1 deck of *Number Sort* Label Cards per group (page A-19)
- Optional: 1 *Number Sort* Directions per group (page A-20)

Directions

Goal: Guess the two sorting rules.

- Decide which team is Team 1. The other team is Team 2.
- Team 1 looks at the cards and chooses two labels, one for each ring of the diagram. Team 1 places each label facedown beside its correct ring and does not tell Team 2 what the labels are.
- Team 2 needs to guess the labels. It chooses a numbered chip and gives it to Team 1. Team 1 correctly places the numbered chip in the diagram.
- Play continues until Team 2 correctly guesses the labels.
- Team 2 must guess both labels at once. Team 2 should also explain its thinking when it makes a guess.
- If Team 2 guesses one label right and one label wrong, Team 1 just says, "No." Team 1 doesn't give any hints about which guess is right or wrong.
- When Team 2 guesses the right label for each ring, Team 1 shows the labels.
- Have teams change roles and play another game.

Number Sort Label Cards

Less than 10	Greater than ten
Every digit in the ones place is the same	Every digit in the ones place is different
Has a 2 in the tens place	Has a 1 in the tens place
Greater than 20	Less than 20
The digit in the ones place is greater than the digit in the tens place	The digit in the tens place is greater than the digit in the ones place
Odd number	Even number

How It Looks in the Classroom

A first-grade teacher invites the students to sit on the rug around a two-ring Venn diagram drawn on chart paper. One *Number Sort* label card is placed facedown above each ring. The chips, numbered 1–35, are placed faceup to the side of the diagram. The students are familiar with Venn diagrams, having used them to compare and contrast traits of animals, with labels such as *Flies* and *Has two legs*.

The teacher invites Jake to choose a numbered chip and hand it to the teacher, who places it where it belongs in the diagram. This process is completed seven more times, and now the diagram looks like the one shown in Figure 4.2. The teacher traces around Ring A and asks, "How are these numbers all alike?"

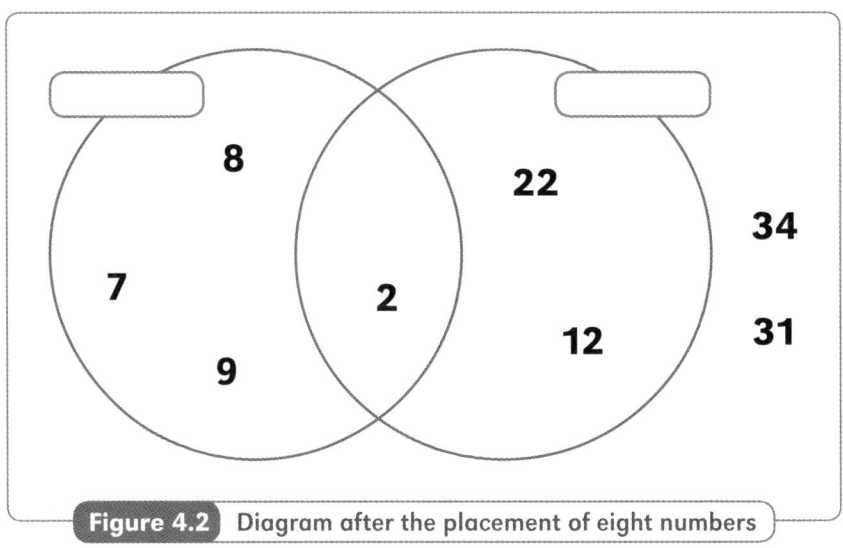

Figure 4.2 Diagram after the placement of eight numbers

Asmina responds, "They're all small."

When asked how small they are, Maribel says, "Less than ten." The teacher has students confirm that this is true as well as check that all the numbers outside Ring A are ten or greater.

The teacher then asks about Ring B, and Tracy volunteers, "Numbers greater than two." Several heads nod and the teacher asks whether all the numbers in Ring B are greater than two.

Many students agree until Russell says, "The two is there. We have to say *two or greater.*"

Once others agree, the teacher says, "OK, all the numbers in Ring B are two or greater. Let's look at the numbers outside of Ring B. Are any of these numbers *two or greater*?" The expressions on some students' faces suggest they recognize that since there are numbers outside of Ring B that also fit this rule, it cannot be the correct rule. A few others do not realize this yet. The teacher knows this is a challenging concept and suggests they place some more numbers.

After a few more turns, Jessica announces, "They all have a two in the ones place; they're all the same."

The teacher asks students to check. "Do all the numbers in Ring B have the same digit in the ones place? Does any number outside of Ring B have that digit in the ones place?" Then she turns over the two label cards and congratulates the students for guessing both labels correctly.

The teacher then displays a list of all the possible labels and asks questions such as *Who can tell me a number that is greater than twenty?* and *Who can tell me a number that has a one in the tens place?* while referring to the labels. Next, she starts a new game, leaving the list of possible labels in view. "This time," she says, "you have to get each label correct at the same time. If you are right about one and not about the other, I'll just say your guess is not right." After this large-group game, students play in groups of four, with two players on each team.

Tips from the Classroom

- If your students are unfamiliar with Venn diagrams, you may wish to take them to an open area such as the gym or blacktop. Make a very large two-ring Venn diagram that they can stand in, and create labels such as *Has a sister, Wearing yellow,* or *Likes to dance.* Invite a few students to individually decide where they should stand in the diagram by asking questions such as *Do you have a sister?* Emphasize that standing outside the rings is as important as standing anywhere inside the rings. (Some teachers use the labels *Is never nice* and *Is never happy* to reinforce this idea.)
- When playing this game, it may be helpful to have the team that chooses the labels place them on a drawing of a two-ring Venn diagram that's hidden behind an upright file folder. That team can hold a chosen numbered chip next to each label to see if it matches and then decide where to place it on the visible diagram.
- You may wish to give guessers a copy of page A-19 so that they can review the label choices as they consider the possibilities.
- Encourage the guessing team to predict where a chosen number will be placed. Making a prediction will help the guessers clarify their intuitions.
- Occasionally suggest that guessers stop after a few numbers have been placed and think about what the labels could be. Such reflection will help the team identify good choices for the next numbered chip they choose.
- Some students may want to guess the labels as soon as one or two numbers are placed, while others may prefer to wait until every number is placed before they make a guess. Over time, assist students in thinking about the best time to make guesses.

What to Look For

- Do students readily identify digits in the tens and ones places?
- Do guessers choose numbers randomly or do they develop a strategy?
- Do some students guess too quickly or guess well after enough information was available to identify the correct labels?
- How do the players who know the labels decide where to place a numbered chip?

Variations

- Limit the game to one ring and one label, with numbers placed inside or outside the ring.
- To encourage strategic thinking, you could limit the number of guesses that may be made per game. You could also specify the amount of numbered chips that must be placed before a team can guess the labels (for those who guess too quickly) or limit the amount of chips that can be placed before the team must identify the labels correctly (to encourage some risk taking and more careful choosing of numbered chips).

> Extend labels and numbers to include three-digit numbers. A possible label could be *The value of the number is greater than 12 tens.*

Exit Question Choices
> You are teaching a friend how to play this game. How will you explain what to think about when you decide where to place a number?
> You are playing a game and this is how the diagram looks. What possible labels would you guess? Why?

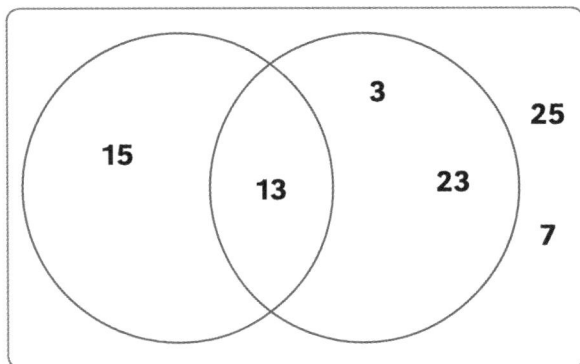

Consider the student response to the second exit question shown in Figure 4.3. The student used the label cards to identify his choices. Note that he chose labels that matched the numbers in each ring; that is, each number inside Ring A fit the labels he chose, and the same is true for the numbers inside Ring B. His work suggests to us that he understood the labels but may not have fully understood Venn diagrams, as he considered the numbers inside the rings but not those outside. For example, all of the numbers in Ring B are indeed odd, but for that to be the correct label, the 7, 15, and 25 would have to be in that ring as well.

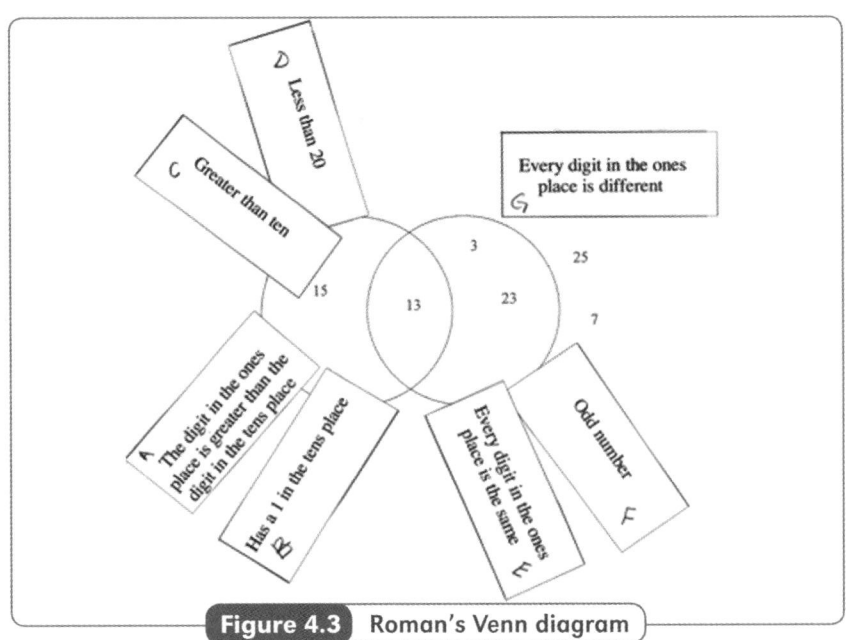

Figure 4.3 Roman's Venn diagram

CHAPTER 4
Base Ten Numeration

Extension

Have students randomly place one numbered chip in each of the four regions in the two-ring diagram. Have them go through the labels and talk about which labels are possible, given the placement of the four numbers. Remind students to look at all four regions to make their decisions. For instance, the two numbers in Ring A may have a two in the tens place, but if there is a number outside of this ring that is also in the twenties, the label would be incorrect. Then have students randomly place another numbered chip and consider what labels are still possible. Or, ask students to place a numbered chip so that a particular label would be eliminated.

Go Number Fish

Why This Game or Puzzle?

The National Council of Teachers of Mathematics, in *Principles and Standards for School Mathematics* (NCTM 2000), identifies representations as essential to supporting students' understanding of mathematical concepts. It is important that students are able to recognize and match the values of representations in the form of standard numeral, number line, number name, and expanded place-value formats.

This game is adapted from the classic card game Go Fish. In this version, players try to make "packs" from three matching representations of a given number. Nonstandard representations, such as *1 ten 14 ones*, are included. Understanding that the value of this combination is twenty-four is essential to understanding standard computational algorithms.

Go Number Fish Cards

12		
24	1 ten 14 ones	
37	3 tens 7 ones	thirty-seven

Math Focus

- Matching different representations of a number
- Understanding that 10 ones is equal to 1 ten

Materials Needed
› 1 file folder per team
› 1 deck of *Go Number Fish* Cards per group (pages A-21–A-23)
› Optional: 1 copy of *Go Number Fish* Directions per group (page A-24)

Directions
Goal: Make the greater number of packs of cards.
› Decide who goes first.
› Mix up the cards. Deal five cards to each team.
› Teams place their cards behind a standing file folder so the other team can't see them.
› Place the other cards facedown in a deck.
› Take turns.
› On each turn, choose a card behind your folder and ask if the other team has a card with the same value. You might ask, "Do you have a card with a value of twelve?"
› The other team must give you any cards it has that match what you asked for, and then you get another turn.
› If the team does not have such a card, it tells you to "go number fish." You fish by taking the top card from the deck.
› If the card has the value you asked for, show it, put it in your hand, and take another turn. If not, your turn is over.
› When you get three cards with the same value, you have made a pack. Place each pack faceup in front of your folder.
› The game ends when a team has no cards left.
› The team with the greater number of packs wins.

How It Looks in the Classroom
A first-grade teacher writes *15* on the board and asks students how else they could show this number, without using addition and subtraction. She asks volunteers to share their ideas by drawing them on the board. Mei draws fifteen small squares and then Drew draws a long rectangle and five small squares to show 1 ten and 5 ones. No other students raise their hands, and after waiting about four seconds, the teacher asks, "Is there anything you could write to show the number?" That question triggers Paul to write *fifteen* and Sophie to write *1 ten and 5 ones*. After an appropriate pause, the teacher asks if anyone could draw a number line. Derek draws a line with the numbers *1–15* with a circle around the 15. The teacher then shows another number line with only the 10 and 20 identified and an arrow at 15 (see Figure 4.4). She asks Amy to count up from 10 to 15 along the line.

The teacher then asks students about the game Go Fish, and they offer the rules they remember. She clarifies the rules and says that this time they will play by matching different representations of the same number to make packs of three.

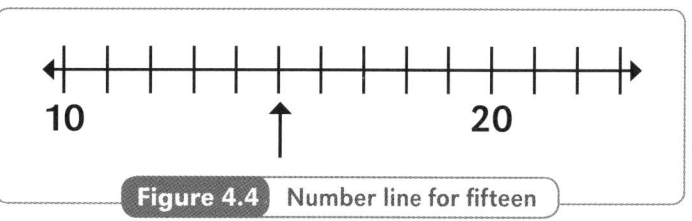

Figure 4.4 Number line for fifteen

Tips from the Classroom

- We found that some students needed to look at a number line with all numbers included to interpret the lines shown on the cards, which identify only the decades.
- Some students will benefit from using a large place-value chart and base ten materials to help them think about standard representations. Have these materials available for easy access during the game.
- Some students may prefer recording numbers in a small place-value chart and then thinking about trades that could be made, similar to what is shown in Figure 4.5.

What to Look For

- What strategies do players use to determine the values of numbers shown in various representations?
- When players ask if the other team has a match for one of their cards, do they identify its value or just show the card?
- Do players find it easier to identify the values of some representations than others?
- Are teams sharing the responsibility for determining the number values?

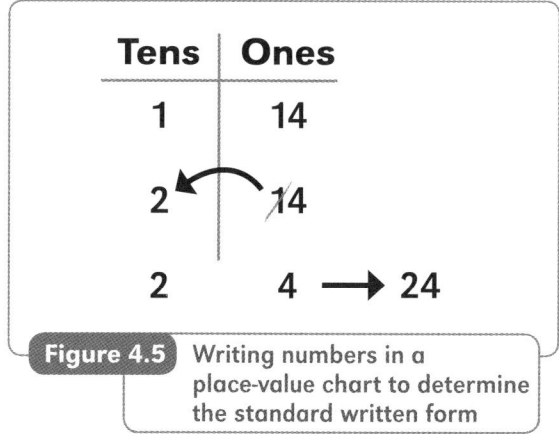

Figure 4.5 Writing numbers in a place-value chart to determine the standard written form

Variations

- Create a similar set of cards for three-digit numbers.
- Eliminate one of the three representations for each number and have players match only pairs.
- Have students create a fourth representation for each number, and require players to get all four matches to make a pack.

Exit Question Choices

- We want to add 18 to the deck. What three cards could you make for this number?
- Which representation (such as a number line, base ten blocks, or the written number) do you find challenging? Why?

Extension

Include place-value riddles in the morning routine such as *There are 13 tens and 14 ones. What number is this?*

The Number Is/What Number Is?

Why This Game or Puzzle?

For this puzzle, each group of students receives a copy of a set of cards, each of which lists an answer and a question. There are two sets of cards provided (see page A-25 and page A-26). The students must arrange the cards in a circle so that the answer to the question on one card is shown on the card that follows it. Further, all of the cards must be included in the circle, and the last card must match the first card.

Some questions have more than one correct response, though only one choice will allow all of the cards to be included in the solution circle. The need to work as a team, with all of the cards, leads students to engage in conversations about how best to work together, which choices to make, and what to do when some cards have been excluded from their solution.

Questions such as *What number is 1 ten less than 7 tens?* are puzzling to many students, which gives them opportunities to build perseverance in solving problems. These questions also provide opportunities to recognize a variety of ways to represent a number, a skill essential to fully understanding regrouping in the traditional addition and subtraction algorithms (Dacey and Collins 2010a and b). Other questions—such as *What is 1 ten more than 32?*—emphasize the structure of our base ten number system because it is not necessary to add using paper and pencil; rather, you can just increase the tens digit by one.

The Number Is/What Number Is? Cards A

The number is 80. What number is 1 ten more than 19?	The number is 29. What number is 7 tens and 3 ones?	The number is 73. What number is 1 ten less than 6 tens?
The number is 50. What number is 1 ten more than 32?	The number is 42. What number is 6 tens and 8 ones?	The number is 68. What number is 1 ten less than 40?
The number is 30. What number is 8 tens and 9 ones?	The number is 89. What number is 1 ten more than 49?	The number is 59. What number is 9 tens and 7 ones?
The number is 97. What number is 1 ten less than 2 tens?	The number is 10. What number is 2 tens and 5 ones?	The number is 25. What number is 10 more than 7 tens?

Math Focus

› Understanding that two-digit numbers are composed of tens and ones or that three-digit numbers are composed of hundreds, tens, and ones

- Applying place-value concepts to identify numbers that are 1 ten more or 1 ten less
- Matching different place-value-based representations of a number
- Understanding that 10 tens is equal to 1 hundred (Deck B only)

Materials Needed

- 1 deck of *The Number Is/What Number Is?* Cards (A or B) per group (page A-25 or A-26)
- Optional: 1 *The Number Is/What Number Is?* Directions per group (page A-27)

Directions

Goal: Place cards so that the number identified on each card answers the question on the card before it.

- Mix up the cards and place them faceup on a table or the floor.
- Choose a card and read its question.
- Find a card with a matching answer.
- Place this card next to the first card.
- Read the question on this second card. Find a card with a matching answer and place it next to the second card.
- Continue to read questions and find answers. Put the cards in a circle so that each question is followed with a correct answer.
- Each card must fit in the circle.

How It Looks in the Classroom

The teacher introduces the puzzle by telling students that she will show them cards that they must put in order so that each question is followed by a correct answer. She then displays the following mini-puzzle (see Figure 4.6) on the interactive whiteboard, so she can move the "cards" as the students direct. She reads the question on Card A, and Nick suggests that the number is fifty. She moves Card C next to Card A and reads the new question, at the bottom of Card C.

Maddie says, "That's twenty-eight," and the teacher places Card D beside Card C. The teacher reads the new question on this card, and Kevin identifies the answer as eighty-two. The teacher moves this card into the last position. The teacher again reads the new question and the students are excited to see that the answer is on the first card (see Figure 4.7). Then the teacher tells them that they are going to solve a puzzle with more cards in groups and that their job is to work together to put *every* card in a circle so that each question is followed by the correct answer.

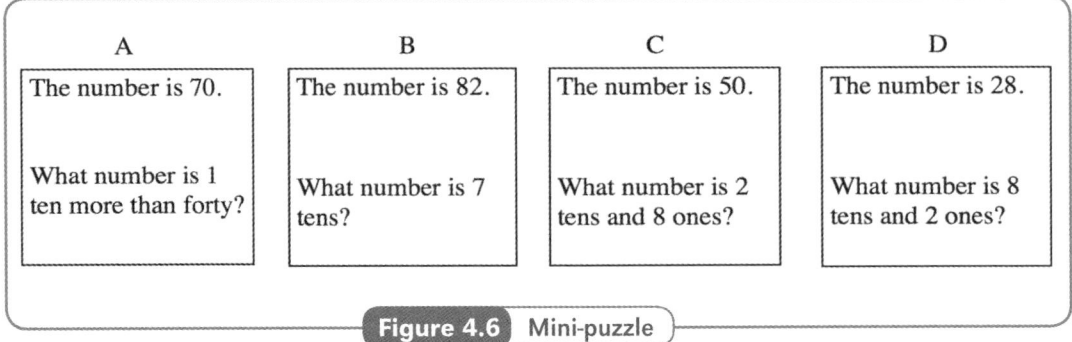

Figure 4.6 Mini-puzzle

Tips from the Classroom

> Printing the cards on sticky notes or putting a magnet on the back of each card and arranging them on a metal cookie sheet may help some students organize their work more easily.
> Some students may find it easier to place the cards in a line. Just remind them that the last card must lead back to the first.
> Having students work in groups of two or three is best.

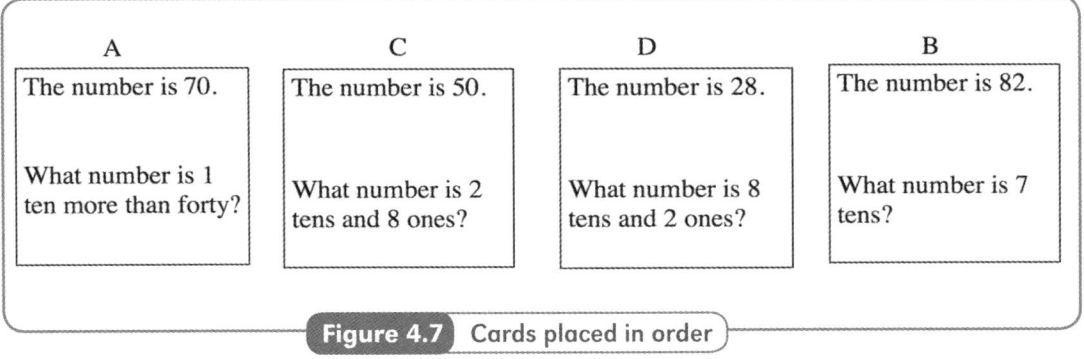

Figure 4.7 Cards placed in order

What to Look For

> How do groups decide how to begin solving the puzzle?
> How do students determine equivalent representations of numbers?
> What do students do when they have yet to find a way to fit all of the cards in the circle?
> What puzzle-solving strategies do you observe that you want students to share with the larger group?

Variations

- Add an "ask a friend" rule, which allows students to ask a classmate working in another group for a hint.
- You can simplify the puzzle by eliminating the last row of pieces. For Deck A, for example, change the last question in the third row from *What number is 9 tens and 7 ones?* to *What number is 8 tens?* so that it will correspond to the first card.

Exit Question Choices

- Write three different questions for the answer 89.
- What number is 1 ten less than 400? How do you know?

Figure 4.8 shows a second grader's response to the first exit question. Note that the student demonstrates an understanding of tens and ones as well as some thinking about the inverse relationship between addition and subtraction.

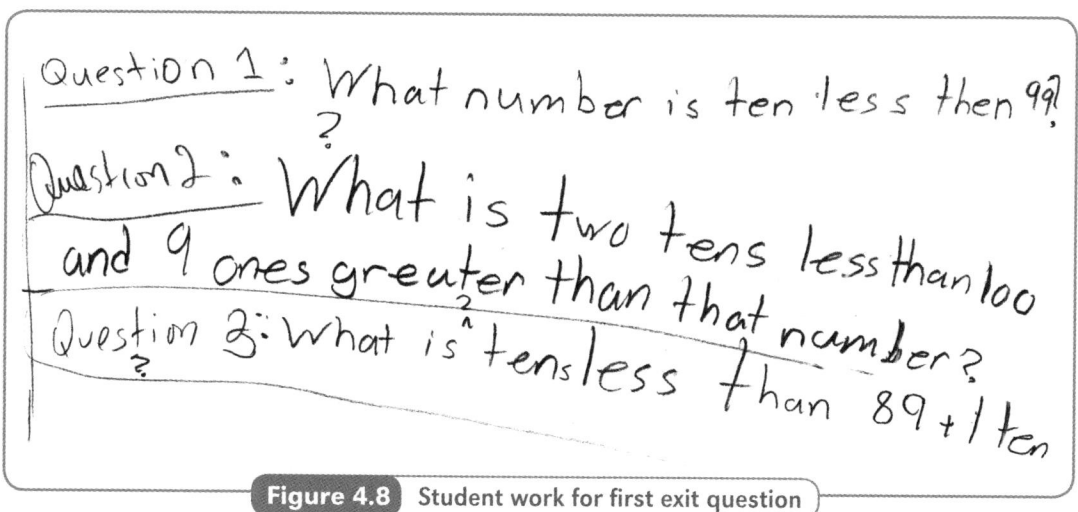

Figure 4.8 Student work for first exit question

Extension

Provide a challenge such as *How many different collections of tens and ones blocks can you make with a value of 51?* or *How many different collections of hundreds and tens pieces can you make with a value of 230?*

This puzzle is adapted from the game *I Have/Who Has?* The puzzle format, however, means that students' turns are never over and conversations are encouraged. You can use cards from preexisting *I Have/Who Has?* games as a puzzle.

Number Touch

Why This Game or Puzzle?

Mathematics instruction that focuses on the magnitude of numbers has a positive impact on student learning (Faulkner 2009). Students need to know the values of the different places. In this game, students identify a number with a given digit and place value, a skill often practiced on worksheets. This game requires this same skill but includes strategy as well. Students need to consider which choice will give them the greatest number of points, as they receive 1 point for finding a number and 1 point for each previously identified number it touches. They may also want to consider the spatial implications of their choices, that is, whether other choices for future play are blocked or not.

Math Focus

- Identifying the value of digits in a two- or three-digit number
- Reading and writing numbers
- Evaluating alternative choices

Materials Needed

- 1 *Number Touch* Game Board per group (page A-28)
- About 10 each of tongue depressor sticks labeled *hundreds*, *tens*, and *ones*, mixed up and placed in an envelope, per group
- 1 deck of *Number Touch* Digit Cards per group (page A-30)
- 1 *Number Touch* Recording Sheet per group (page A-29)
- Optional: 1 *Number Touch* Directions per group (page A-31)

Number Touch Game Board

451	270	382	95	247	81
302	194	45	148	536	253
36	29	468	379	50	79
527	310	73	92	487	14
104	265	90	536	61	128
49	386	58	103	279	75
413	97	152	258	304	580
260	74	521	85	410	93

Directions

Goal: Earn the greater number of points by finding numbers on the game board and earning points for those numbers and any number they touch.

- Use rock-paper-scissors to decide which team goes first. The winner goes *second*.

- Place the cards facedown. Move them around so they are shuffled.
- Teams take turns. On each turn, choose a card and a stick. Look for a number on the board with the digit on the card in the place on the stick. Mark it with an *X*. This number may not be marked again. For example, if you draw the card with the 2 and a stick labeled *tens*, you should look for a number with a 2 in the tens place.
- You get 1 point for finding a correct unmarked number on the board and 1 point for each other *marked* number next to it (in a row, column, or diagonal). If you could mark 148 on the mini-board shown below, you would receive 3 points: 1 point for 148, 1 for touching 536, and 1 for touching 247.
- If you cannot mark a card, your turn is over.

382	95	2̶4̶7̶
45	148	5̶3̶6̶
468	379	50

- Each team gets eight turns.
- Write your score on the recording sheet. After the first turn, be sure to add your old score to your new one before you write. The team with the greater score at the end wins.

How It Looks in the Classroom

Before introducing the game, one second-grade teacher wants to stimulate students' thinking about how many different choices they will have in the beginning of the game. She tells the students that they are going to do a "quick look" activity. She randomly chooses a digit card and a place-value stick. She looks at them and says, "I am going to show you a lot of numbers. I want you to look for those that have a 3 in the hundreds place and write down those numbers. Pay attention; you will only get a quick look." She then briefly displays the *Number Touch* Game Board (page A-28).

Some students record each number as they find it; others look and then record two or three numbers at once. They all groan when the display is turned off. The teacher tells them to put their pencils down and then turns the display back on. She has volunteers name a number they recorded and come up and cross it off. Once everyone's recorded numbers are identified, they look for more numbers that fit the rule.

Next the teacher displays a fresh game board and explains the rules of the game. The teacher draws a digit card and a place-value stick, looks at them, and says, "The students on the right will be one team. This team should find a number with a 4 in the hundreds place." The team identifies 487 and the teacher crosses it out and tells the students, "Once a number is used, we cross it out to show that it can't be used again. Your team gets 1 point."

The teacher chooses another digit card and place-value stick, looks at them, and turns to the students on the left side and says, "Now it is your turn. You will get 1 point for finding a number with a 2 in the ones place. You will also get a point if your number touches 487." After three more rounds, she asks students to explain the rules of the game to her before having them play in teams.

Tips from the Classroom

- If you decide to use the quick look activity, partner students with different abilities and differentiate roles. For example, a student may not have a strong memory but may be good at scanning data visually. Have that student say the numbers for his or her partner to record.
- When playing the game, some students may take a long time to make a choice. You may want to include a timer to limit how long opponents have to wait for their turns.
- During the field testing, one group of students wanted to consider two-digit numbers as numbers having a 0 in the hundreds place. This idea was challenging for us, as their argument was that there were zero hundreds. While it is true that those numbers do not have any hundreds, we do not write *0* as the first digit in a number. Because writing such digits when subtracting is a common error students make when using traditional algorithms, we decided to explain that they had to see the digit in the number on the board in order to cross off the number.

What to Look For

- Do students correctly identify the place values?
- What language do players use when talking about which choice to make? Do they refer to the place values or talk about the digits as if they were all ones?
- Do students explore several possibilities or choose a number quickly?

Variations

- Make a board with only two-digit numbers and eliminate the hundreds sticks.
- To encourage engagement of all players during the entire game, add the rule that if opponents find a choice that is worth more points, they get 1 point.
- For students who can add numbers with five-digit sums, eliminate the place-value sticks and just use the digit cards. Change the rules so that rather than receiving points for identifying a number and touching marked-off numbers, they find a number with that digit and receive points equal to its value. For example, if 9 is the digit and the team chooses the number 194, the team gets 90 points.

Exit Question Choices

› You draw a card with a 2 and a stick labeled *tens*. Which of the following numbers could you *not* cross off if the game board showed the numbers 22, 202, 21, and 221? Why?

› You crossed off the number 382 on the game board. What might have been on your card and stick? What other possibilities can you name?

Extension

Have students write in their journals or talk about where in their daily lives they most often see numbers that have only ones; have tens and ones; or have hundreds, tens, and ones.

Online Games and Apps

Technology can provide students with the opportunity to visualize base ten numeration, actively engage in the pattern of the count, and incorporate critical thinking with learning about number sense. Moving beyond number recognition, such games require students to apply their understanding of the number system rather than to memorize a number sequence or a list of place-value names by rote. Many such games also give instant feedback to the player, providing information that can support students' conceptual understanding.

- A free online game titled Math Match, found on the Fuel the Brain website at http://www.fuelthebrain.com/games/math-match/, allows students to play a concentration game with various levels of difficulty. The first level requires players to match one- and two-digit numbers with their physical representations using base ten blocks. A second level incorporates the word representations for the numbers. The game includes twenty overall levels, allowing for differentiation and continued challenge.

- The Place Value Game for Kids, available for free on the Kids Math Games Online website at http://www.kidsmathgamesonline.com/numbers/placevalues.html, may be considered more of a puzzle than a game. The puzzle provides clues to partners, who must then create numbers (of at least three digits) that meet the criteria provided in the clues. Thought-provoking clues such as "Find a number between 425 and 450" and questions such as "Using the digits 2,

3, and 6 just once each, what is the greatest number you can make?" require deductive reasoning as well as mathematical understanding. Partners are given digits to use to create the number by placing them in a place-value chart. When the program checks the number for accuracy, it provides feedback in the form of a written description of the number that was created.

- Shark Numbers, a free game from ICT Games at http://www.ictgames.com/sharkNumbers_v2.html, allows players to choose the manner in which the numbers are modeled, either with base ten blocks or with stacks of cups. A number is then shown with the chosen model, as three possible numerical representations are moving along the bottom of the screen. If partners choose the correct number, a dolphin moves across the screen, but if they choose an incorrect number, the shark takes a bite out of the boat. Number ranges allow for differentiation, as students may choose to play with numbers up to 29, 59, 99, or 999.

CHAPTER 5

Addition

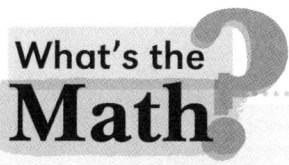

What's the Math?

Over time, as students develop more sophisticated strategies, addition develops as an extension of counting (Eisenhardt et al. 2014). Initially students model addition by counting each set, combining the sets, and counting *all* to determine the total. Counting *on* is an important developmental milestone. Once they've reached it, students can make both sets but count on from the first set, instead of combining the sets and counting all. Later, students can apply this concept abstractly. That is, to find $9 + 3$, a child might count on, "Ten, eleven, twelve," while keeping track that he or she has counted three numbers up from nine.

In kindergarten students generally explore addition with sums to ten and develop fluency with facts to five. Fluency involves four basic components: flexibility, appropriate strategy use, efficiency, and accuracy (Kling and Bay-Williams 2014). In first grade, students usually explore addition with sums to 100 and obtain mastery with sums to 20. Students are likely to develop fluency with sums to 100 and extend the exploration of addition to sums of 1,000 in second grade.

Strategies for finding sums might involve direct modeling, applying properties, adding by place, or making a connection to an equivalent but easier example (Dacey 2014). Such approaches develop the conceptual underpinnings of common addition algorithms, and it's necessary for students to receive significant exposure in order to internalize their use. Games and puzzles in this chapter help students become fluent with addition while developing conceptual understanding and strategic thinking.

# Make a Pair

Why This Game or Puzzle?

Five is an important anchor number for young learners. In this game, players look for visual models of two numbers that sum to five. The numbers are represented on five-frames, the top row of a ten-frame. As students progress, they can play the same game with ten-frames. (See the "Variations" section and page A-33.) Use of representations such as these frames supports students' ability to recognize small sets of numbers without counting, a capability referred to as *subitizing* (Clements 1999).

Starting with a five-frame, instead of a ten-frame, is important. Building confidence with these representations supports later work with ten-frames, as students can recognize how many more than five (or less than ten) there are. This recognition can support both addition and subtraction, as shown in Figure 5.1.

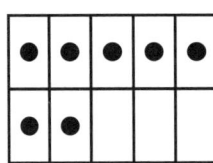

I have two more than five. I have seven. I need three more to make ten.

Figure 5.1 Five-frame thinking supports working with ten-frames and with making connections between addition and subtraction.

Math Focus

› Modeling addition
› Adding with sums to five

Materials Needed

› 1 deck of *Make a Pair* Five-Frame Cards, made from 2 copies of page A-32 per group
› Optional: 5 counters per team
› Optional: 1 *Make a Pair* Directions per group (page A-34)

Directions

Goal: Be the first team to find five pairs of five-frame cards with a total of five dots.

- Mix up the cards. Put them facedown in a pile.
- Each team takes three cards from the pile.
- Turn over the top card of the pile and place it faceup beside the pile.
- Take turns.
- On each turn you can do one of three things:
    1. Put two of your cards down, if you think they have a total of five dots. If the other team agrees, you have a pair.
    2. Pick up the card that is faceup and put it down along with one of your cards, if you think the total number of dots is five. If the other team agrees, you have a pair.
    3. Take the top card of the pile and see if you can make a pair. If the other team agrees, you have a pair. If you do not have a pair, keep the card for use another time.
- After each turn, if you do not have at least three cards, pick up cards from the pile until you have three cards in your hand. Also, turn over a new card from the pile, if one is not shown faceup already.
- The game ends when a team makes five pairs. The first team with five pairs wins.

How It Looks in the Classroom

A kindergarten teacher gives each student a set of five-frames that represent the numbers zero through five. She displays the card representing three and says, "Hold up this card and check to see if your neighbors are holding the same card as you are." She repeats this process for two and then five. Then she holds up the frame for four and asks, "What number does this show? How do you know?"

Kim Su answers four and counts aloud as she touches each of the dots. Ben explains that he just sees it, and Jeri reports, "It's four because I know it is one less than five dots."

Next the teacher holds up the frame for four and the frame for one. She asks, "What is the total number of dots on these cards? Talk with your neighbor about how you know."

As she observes her students, she notes that most of them put the two frames together and count all of the dots. She sees a few students count on from the four. She has students who used each strategy explain their thinking to the class. Then Carista says, "I did something different. I thought about the one dot moving over to the other card and that would be five dots."

After each strategy is shown, the teacher asks another student to restate it. Then she writes the corresponding number sentence on chart paper. As she represents Carista's idea, she says, "So the four dots together with the one dot is equal to five dots."

The teacher follows the same process with the frames that show two and three. After discussing this example, she shows the frames for four and two and asks, "Do these two frames make a total of five dots?"

Huey tells the class it would be six, not five. The teacher is pleased to see Anna build on an idea shared earlier when she says, "We could move one dot here to make a total of five dots. So, two dots are too many."

As the teacher knows her students can find zero challenging, she holds up the card with no dots and asks, "Which card can we pair with this card to make a total of five dots?" Most students recognize that the card with five dots is needed. Then she holds up the card with five dots and says, "Which card would we pair with this one to make a total of five dots?"

Nelson quickly explains that another card isn't needed as there are already five dots. The teacher responds, "But, if we want to make a pair?" Nelson then agrees that the card with zero dots could be used. The teacher writes the corresponding number sentence on chart paper as she says, "So the five dots together with zero dots is equal to five dots."

The teacher then explains the rules of the game and students play in teams.

Tips from the Classroom

> Many students could recognize five dots on a five-frame, though they needed to be reminded that the match would be a frame with zero dots.

> When students counted their pairs, they did not always know that it takes two cards to make a set. Some students would benefit from having a way to keep track of their pairs, such as a graphic organizer with five places to put their pairs as they collected them.

> If counters are available, students can place one on each dot of the two cards and then count the manipulatives to make sure the total is five.

What to Look For

> Do students recognize the number represented on a frame without counting?
> How do students find two frames that they think have a total of five dots?
> How do teams check that a pair has been made correctly?
> Are students ready to explore ten-frames?

Variations

> Replace the five-frames with four sets of number cards labeled *0–5*.
> Have students play with ten-frames (make four copies of page A-33).
> Have students play with either the five-frames or the ten-frames, but play the game like a concentration game. They place all of the frames facedown between the teams.

On each turn, a team turns over two cards and keeps them if they show a total of five (or ten) dots. If they do not, the team turns the cards back over. For this game, you may want to limit the cards to those made from one copy of page A-32 or two copies of page A-33.

Exit Question Choices

▸ You have this card: [• • • □ □]

How many dots should be on another card to make a total of five dots?
▸ What are two ways you could find the total number of dots on two cards?
▸ How do you know when two cards will not have a total of five dots?

Extension

Have students identify pairs of numbers with a sum of five. Note that some students might list pairs of numbers, while some might write addition sentences. Responses may vary from identifying one such pair to identifying all six.

Equal Values

Why This Game or Puzzle?

This game requires students to recognize that two expressions have the same value. Players might do so by modeling or counting to find sums or by applying fact knowledge. Students might also realize, or discover through playing the game, that addends written in a different order have the same sum (commutative property of addition). They might also note relationships among expressions. For example, a team might decide that $3 + 4 = 2 + 5$ because in the second expression, one addend is one less and the other is one more than the addends in the first expression.

After finding a match, students must record the expression on each card on either side of an equal sign. Many students misinterpret the equal sign (Ginsburg and Ertle 2008), often interpreting it to mean *write your answer here*. Students may also assume that only one number can be written to the right of the sign. Recording pairs of expressions provides students with counterexamples to these assumptions and also provides a list from which patterns and generalizations can be discussed.

Math Focus
- Finding sums of single-digit numbers
- Applying the commutative property of addition
- Recording expressions with equal values

Materials Needed
- 1 deck of *Equal Values* Cards per group (pages A-35–A-37)
- 1 *Equal Values* Recording Sheet per team (page A-38)
- Optional: 1 *Equal Values* Directions per group (page A-39)

Directions
Goal: Get the most pairs of cards that have equal values.
- Give each team a card with the equal sign.
- Shuffle the remaining cards. Deal each team four cards faceup for all to see. Put the other cards facedown in a pile.
- Decide which team goes first.
- On each turn, you can do one of three things:
  1. Find two of your cards that have an equal value. Set this pair beside you. Replace them with two cards from the top of the pile.
  2. Trade one of your cards with one of the other team's cards when that lets you make a pair. Set this pair beside you. Replace your card with a card from the top of the pile.
  3. Draw a card from the top of the deck and add it to your cards.
- When a team makes a pair, both teams must agree that the sums are equal and then the team that made the pair must record the expressions on its recording sheet.
- If no cards are left in the pile, you can still have a turn, but you can't take a card from the pile.
- The game ends when no team can make another pair.
- The team with more pairs wins.

How It Looks in the Classroom
A first-grade teacher uses a number scale to motivate a conversation about equality. He likes to use balance as a way to think about equality, as students understand that one or more objects could be on either side. He puts a marker on 8 on the left side of the scale and asks students to talk with their neighbors about two numbers they could mark on the right side to balance the scale. The students identify 4 and 4 first and then someone suggests the numbers 3 and 5.

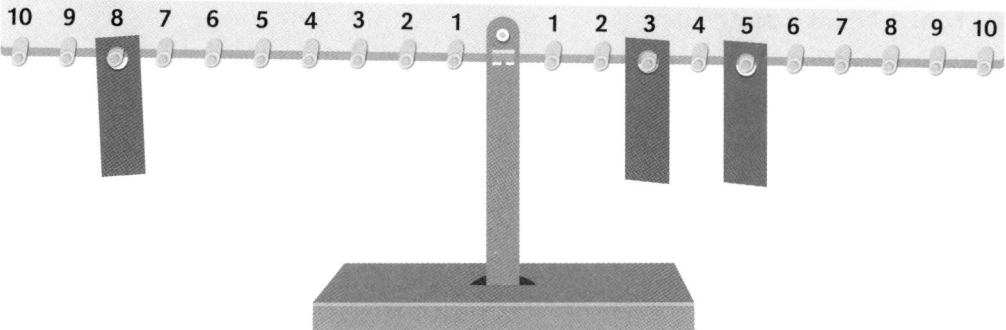

Once the class confirms the scale is balanced, the teacher removes these markers and then places markers on the 2 and the 4 on the right side and asks students to identify two numbers to mark on the other side. Students suggest placing two markers on 3 or placing a marker on 1 and another on 5. The teacher records the two number sentences: *3 + 3 = 2 + 4* and *1 + 5 = 2 + 4*. He points to the equal signs and asks, "What do these signs tell us?"

Taz says, "The sums are equal."

Jodi responds, "That they are the same on each side."

Next the teacher writes two number sentences on chart paper and asks students to decide if each sentence is true or false: *7 + 1 = 4 + 6*; *4 + 5 = 3 + 6*. He tells them he wants them to think alone first and that he will tell them when it is time to talk with their neighbors about their thinking. After some time to think and share in pairs, he calls them back to the larger group. All agree that the first sentence is false, and many give statements such as "Ten is more than eight" or "Eight is less than ten" as their reasons. Nadia suggests she didn't have to compute each side of the equation when she says, "I know 6 plus 4 is 10, and 7 plus 1 is less than that."

For the second statement, Clarissa says, "Well, 4 plus 5 is 9 and 3 plus 6 is 9, and 9 equals 9, so it's true." The teacher is pleased with the deductive reasoning and that the student knew to add both numbers on each side, rather than compare the sum of 9 on the left to the first number, 3, on the right.

Jared says, "It's really cool that four is one more than three, and five is one less than six." The teacher notices that he is thinking more relationally than many of the students. He makes a note to come back to this idea after they have played the game, hoping more students will be able to consider this thinking.

To introduce the game, the teacher deals cards under the classroom projection device and invites two teams to play while others watch and listen. To keep the observing students involved, he occasionally stops and asks questions such as *Is there a different move you would have made? Why do you think this team was unsure about whether or not to trade a card?* and *What card might this team like to get when it takes one from the deck?*

Tips from the Classroom

- At first we were uncertain as to whether students should stop playing to record their matching expressions. Through field testing we found that it did slow down the game, but that the slower pace encouraged students to pay attention to whether or not the expressions were equivalent.
- A game may require two recording sheets.
- We found the equal signs helped students compare the expressions and understand what to record. You may wish to make the equal signs on paper stock of a different color or cut off the white space around them, so they don't get lost among the other cards.
- Have manipulatives available to help students prove or disprove that the expressions are of equal value.

What to Look For

- What strategies do students use to find sums? Do you see any evidence that students are thinking about relationships among the addends?
- Do students recognize examples of the commutative property?
- What language do students use when deciding whether two expressions are equal or not?
- Do players find the values of their opponents' cards before making a trade?

Variations

- Have students use only the cards on page A-35, with sums to five, and deal only two cards to each team. Or, have students use the cards on pages A-36 and A-37, and deal three cards to each team.
- Create cards with two- or three-digit numbers. To support mental math strategies, emphasize the commutative property, for example, 42 + 36 and 36 + 42. Also include expressions that encourage students to think about relationships among the numbers, such as 89 + 45 and 90 + 44.
- Have students play with subtraction expressions and then with both addition and subtraction expressions, allowing students to note inverse relationships, such as 3 + 4 = 7 and 7 − 4 = 3.

Exit Question Choices

- How do you know that 4 + 5 = 8 + 1?
- Is it true that 8 + 7 = 7 + 7? Explain your thinking.
- What two expressions can you write that are equal to 3 + 6?

Figure 5.2 Student response to first exit card

A response to the first exit question is shown in Figure 5.2. Note that the student includes a visual representation as well as a written explanation.

Extension

Post several expressions throughout the room and have a scavenger hunt. Teams can use the *Equal Values* Recording Sheet (page A-38) to record the expressions they find with the same value.

This game is adapted from an unpublished game created by first-grade teacher Suzanne Lak.

Why This Game or Puzzle?

Triangle Totals puzzles are a classic activity that requires puzzlers to arrange a given set of numbers in a triangle so that the sum of the numbers on each side of the triangle is the same. Most students will use the guess-and-check strategy to arrange the numbers. Strategic guessers will know not to put, for example, the three greatest numbers on one side. Some solvers may think about how they can apply what they learned when solving one puzzle to find the solution of another one.

Solving these puzzles can be challenging and requires perseverance. Persevering in solving problems is the first standard for mathematical practice listed in the Common Core State Standards (NGA Center for Best Practices and CCSSO 2010). Though students will begin with a random approach, with encouragement, many learn to use an incorrect guess to inform future guesses.

Math Focus

- Finding totals of three numbers
- Understanding that you can add three numbers in any order
- Guessing and checking to solve problems

Materials Needed

- 6 chips numbered *1–6* or *4–9* per team
- 1 *Triangle Totals* Puzzles Sheet (A or B) per team (page A-40 or A-41)
- Optional: 1 *Triangle Totals* Directions per team (page A-42)

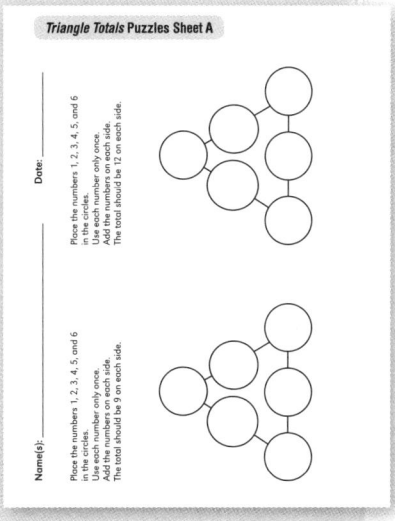

> **Directions**
> Goal: Place the numbers in the triangle so that when you add the three numbers on each side of the triangle, you get the given total.
> - Place each of the given numbers on the triangle.
> - Add the numbers on each side.
> - Check to see if each side matches the given total.
> - If not, try again.
> - Check your solution with another team.

How It Looks in the Classroom

A first-grade teacher introduces a *Triangle Totals* puzzle by modeling some ways to think about how to solve it. She displays a sample puzzle along with six chips, numbered *1–6*, and tells them that each side of the triangle must be equal to ten. (Figure 5.3 shows the puzzle with its solution.) First she places all six numbers randomly and asks students if it could be the solution. Students talk in pairs, find the totals, and decide this is not the solution. The teacher notices that some of the students find the total of all three sides, even after identifying one of the totals is greater than ten. Several students share their thinking as to why this solution is not correct. Then Sonia says, "It can't be right because one side has the 5 and the 6, and that is already more than ten." The teacher wants to emphasize this type of thinking, so she asks another student to restate what Sonia said.

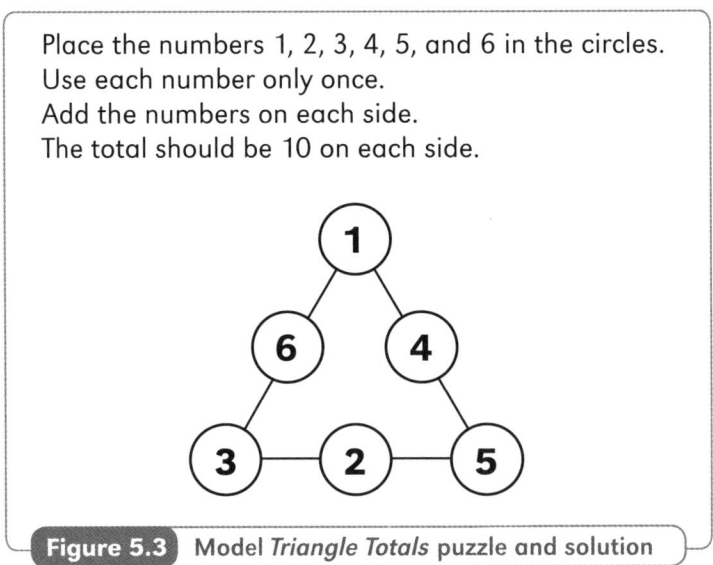

Figure 5.3 Model *Triangle Totals* puzzle and solution

To further illustrate how the possibilities can be limited, the teacher places the numbers 6 and 2 on the same side and asks what students think. They recognize that two more are needed to make a sum of ten, and the 2 has already been used. "Oh," the teacher says, "sometimes we know right away that a number needs to be moved."

To be sure the students have energy for solving on their own, the teacher places the numbers 1, 3, and 5 in the corners, knowing this placement will lead to a solution. She points to the side with the 1 and the 5 and asks, "What sum do we have now? What number is needed?" She continues this process until the puzzle is completed and then tells students to check that the total is the same on all sides. Lastly, she asks, "How did you find one plus four plus five?"

Carolina says, "I added one plus four to get five. Then five plus five is ten."

The teacher asks if they could start at the bottom and add five plus four plus one.

Adam says, "Yes, but it is not as easy."

The teacher then distributes numbered chips and copies of the puzzle for pairs of students to solve.

Tips from the Classroom

> Consider giving some students their own copy of the puzzles and a set of chips. We found that some students do best solving the puzzle alone and others benefit from talking with a partner. Others seem to prefer to work alone for a bit and then check with a neighbor.

> As students work, you may wish to prod their thinking by asking questions such as, *Are there different numbers you could use in the corners?* and *Is there a way to use what you have here to help you decide what to check next?*

> A few students moved all their numbered chips off the puzzle to record their answer and then couldn't remember which number went where. You may want to suggest removing one numbered chip and recording that number before removing another chip.

What to Look For

> Do students appear to guess randomly or organize their guesses in some way?
> What evidence do you observe that students are thinking strategically?
> Do you observe a conversation or a decision-making process that you think should be shared with others?
> What strategies are students using to find the total of three numbers?

Variations

> For extra challenge, don't identify the target sums, but just the numbers to be used, and have puzzlers try to find different arrangements with the same sum on each side.
> Provide support by including partial solutions, perhaps two corner numbers, within the first puzzle.
> Allow solvers to ask a peer if the chosen corner numbers will work.
> You may wish students to record the three related number sentences for each puzzle.

Exit Question Choices

> You are using the numbers 1, 2, 3, 4, 5, and 6 and want a sum of 11 on each side. Where would you write the other numbers?

> You are using the numbers 1, 2, 3, 4, 5, and 6 and are trying to have the same total on each side. Can these be the corner numbers? Why?

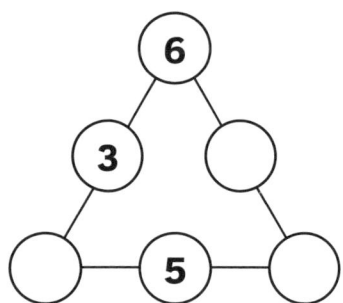

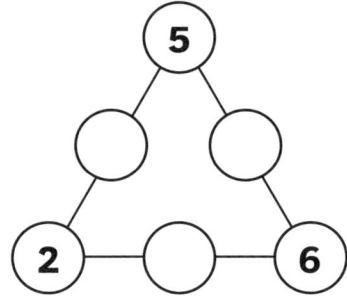

Extension

Post the *Square Totals* puzzle shown in Figure 5.4 and invite interested students to explore it.

Use the numbers 1, 2, 3, 4, 5, 6, 7, 8, and 9. Use each number only once. The 5 has already been placed for you.

How can you place the other numbers to get the same sum in each row, column, and diagonal?

	5	

Possible answer:

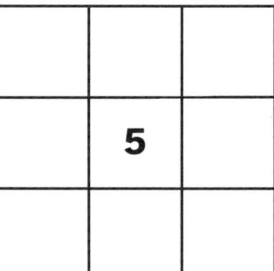

Figure 5.4 *Square Totals* puzzle

CHAPTER 5
Addition

 **Yahoo! 100**

Why This Game or Puzzle?
There are a variety of strategies for finding sums, including counting on, adding one number to the other, in parts, and adding both numbers by place. When students are first learning such strategies, they may make a variety of errors, particularly when regrouping is involved. Familiar contexts can support the development of adding by place or columns and lessen errors (López Fernández and Velázquez Estrella 2011). We believe that games can provide such a context.

In this game, players turn over cards that show single-digit numbers or multiples of ten. They turn over as many numbers as they wish, adding as they do so. Their goal is to get the greatest number of cards with a sum less than or equal to one hundred. One hundred is an important anchor number, and being fluent with sums to one hundred is particularly valuable when mentally computing change from a dollar. As such, players receive 10 bonus points when they reach a sum of one hundred exactly, though it is a rare occurrence. Students must write a number sentence to represent each turn as well as record their score.

Math Focus
- Adding one- and two-digit numbers
- Comparing numbers

Materials Needed
- 1 deck of *Yahoo! 100* Cards, made from 3 copies of page A-43, per group
- 1 *Yahoo! 100* Recording Sheet per team (page A-44)
- Optional: 1 *Yahoo! 100* Directions per group (page A-45)

Directions
Goal: Earn the greater amount of points by collecting numbers to make sets with totals less than or equal to one hundred.
- Put the cards facedown between the teams. Spread them out and mix them up to shuffle them. Leave them spread out for the game.
- Take turns.

> - Begin each turn by turning over two cards. Add the numbers. If you choose to turn over another card, you *must* also add that number to your total.
> - Decide when to stop turning over a card and adding the number to your total.
> - If your total is less than one hundred when you stop, you get 1 point for each card you used. If it is equal to one hundred, you get 1 point for each card and 10 bonus points. If it is greater than one hundred, you get 0 points.
> - Record a number sentence for this turn, the points it is worth, and your total points for the game.
> - Put the cards back facedown and mix them all up again before the other team takes its turn.
> - The team with the greater total score after six rounds wins.

How It Looks in the Classroom

The teacher gathers the game cards near the projection device to first use in a warm-up activity. She displays the cards showing 30 and 6 and asks students to tell the total and explain how they found it. Rosie identifies thirty-six and explains that she counted on six from thirty. Horatio says, "I think of our base ten charts. Three tens and six ones show thirty-six."

Luanne asks if she can come up to the front of the room to show her thinking with the cards. She then takes the 6 and moves it on top of the 0 in the 30 (see Figure 5.5) and says, "I see three tens and six ones, and I know that makes thirty-six." The teacher asks Rowen to come up and use Luanne's thinking for finding twenty plus nine.

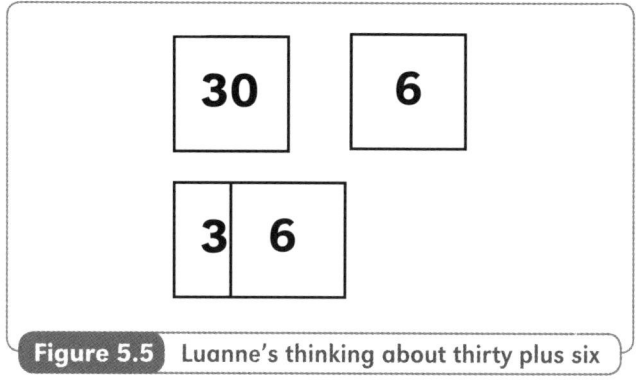

Figure 5.5 Luanne's thinking about thirty plus six

Next the teacher displays the cards showing 50 and 20 and asks students to find their total. The class identifies seventy as the answer, and again, a few students explain their thinking. Then the teacher says, "When we have five tens," pointing to the 50, "and two tens," pointing to the 20, "we have a total of seven tens, or seventy," displaying the card with 70 on it. She turns over the 4 card, and students agree the total is now seventy-four. Then she turns over the 6 card and asks what the total is now. Mark comes up to the front of the room and puts the 4 and 6 together and the 20 and 50 together. He says, "Four and six is ten, so we have ten more than seventy, and that is eighty."

The teacher then demonstrates this game to the whole class. Under the projection device, she shows the cards facedown and then she models how to mix them up by moving them around on the table. She tells the students that their task is to turn over as many cards as they want, to get a total less than or equal to one hundred. She also tells them that they will get 1 point for each card they use and a bonus of 10 points if their total is exactly one hundred. She invites two students from one side of the room to come up, each turn over a card, and find the total of the numbers shown. Once they announce the total of fifty-seven, the teacher says, "You have 2 points because you have a total less than one hundred with two cards. Do you want to turn over another card?" They nod their heads and wiggle with excitement as they turn over another card, which shows 20. The two students visibly relax and they decide not to turn over another card. The teacher tells the two players to record, on the board, a number sentence for their turn and the points they earned. Then she instructs them to place their cards facedown and mix them up with the other cards.

The teacher then invites two students from the other side of the room to each turn over a card. The same process is repeated a few times, as the teacher wants to make sure students understand how to play the game before they begin playing in groups.

Tips from the Classroom

> It helped some students to line up one card below another when finding sums.
> Have hundreds charts and base ten materials available for players' use.
> Students often rely on paper-and-pencil techniques to find sums, whether they need to or not. You may want to encourage all students to try mental computation strategies. Language can be helpful, for example, thinking about five tens and three tens instead of fifty and thirty.

What to Look For

> What strategies do students use to find sums? How do students apply what they know about place value?
> What, if any, misconceptions do you observe?
> Do students make reasonable choices about whether or not to turn over another card?

Variations

> Add cards with multiples of one hundred and have students play *Yahoo! 1,000*.
> Don't have players mix the cards up after a turn; rather, have them put the cards facedown, with both teams able to see where they have been placed. This change will make memory more important and allow for more strategic moves.

› Include a few wild cards that may represent any other card in the deck to encourage students to think about how they might, if possible, make a sum of one hundred from their current total.

Exit Question Choices
› You have turned over the following cards: 4, 20, 50, and 8. What is the total? Tell how you know.
› It is your turn. You have four cards with a sum of seventy-five. Should you turn over another card? Why?

Extension
Challenge students to spend a few minutes writing addition sentences equal to one hundred. One student's work is shown in Figure 5.6. What feedback might you give this student?

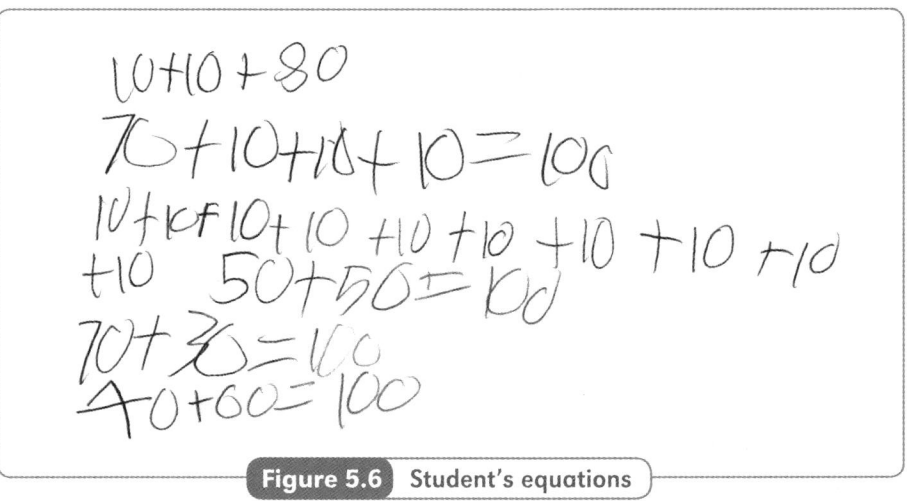

Figure 5.6 Student's equations

 **On Target**

Why This Game or Puzzle?
Number sense includes being able to estimate the results of computation and to recognize that an answer seems about right or that it doesn't make sense. Students need opportunities to develop this aspect of number sense in a way that seems relevant and integrated into their thinking, rather than merely a reminder to estimate to check their sums or differences.

In this game, teams roll dice to form a three-digit target number and then organize a set of six digits to create two three-digit numbers with a sum as close as possible to that target. While deciding where to place the digits, players use estimation to help them make good choices. *On Target* also provides practice in finding exact sums and comparing numbers. As students unpack their thinking about the computational decisions they are making, they are engaging in mathematical thinking and reasoning (White and Daukus 2012).

Math Focus
- Estimating sums
- Adding three-digit numbers
- Comparing numbers
- Recognizing patterns

Materials Needed
- 3 dice per group
- 1 deck of playing cards, with face cards removed, per group
- 1 *On Target* Recording Sheet per team (page A-46)
- Optional: 1 *On Target* Directions per group (page A-47)

Directions
Goal: Place digits in a number sentence to get as close as possible to a given sum.
- Mix up the cards.
- Give each team six cards.
- Roll the three dice, one at a time. The first one rolled shows hundreds, the second shows tens, and the third shows ones. This is the target sum. Write it on your recording sheets.
- Each team places its cards to form two three-digit numbers, writes the numbers on its recording sheet, and records the sum.
- Teams compare the sums. Circle the one that is closer to the target sum.
- Play five rounds. The team with more sums closer to the target numbers wins.

How It Looks in the Classroom

A second-grade teacher introduces this game by presenting the task shown in Figure 5.7. She encourages the students to write the digits on small slips of paper so they can move the digits around. She tells the students that they will have a few minutes to try different ways to arrange the numbers. She instructs them to work with partners and to record what they find. Then she says, "Find as many possibilities as you can."

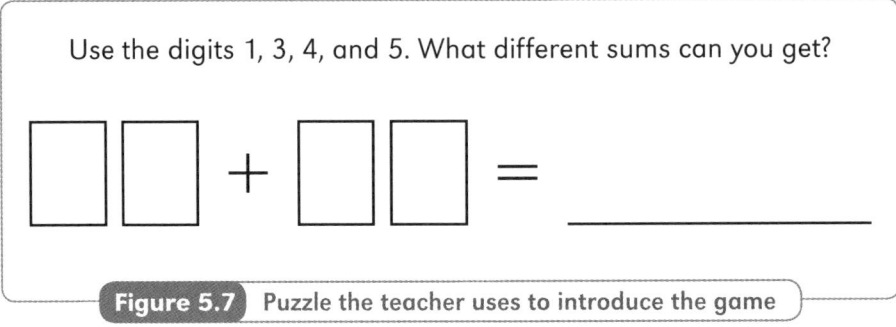

Figure 5.7 Puzzle the teacher uses to introduce the game

After an appropriate amount of time, the teacher says, "We will go around the room, and each set of partners will share an answer they found. If you found a sum that has not already been shared, be sure to tell us that idea." She records all of the number sentences they share:

13 + 45 = 58; 43 + 15 = 58; 15 + 43 = 58
34 + 12 = 46
45 + 31 = 76; 41 + 35 = 76
41 + 53 = 94
14 + 53 = 67; 54 + 13 = 67

"What do you notice about our list?" the teacher asks. Students give a variety of responses, such as "Some of the answers are the same," "There are lots of different sums," and "There are different ways to get the same sum."

To follow up on their thinking, she points to the sum of 46 and asks if they can find another way to get this sum. Chandra says, "The opposite will always work, and so twelve plus thirty-four equals forty-six."

To better prepare them for the game, the teacher asks, "If you were trying to get a sum close to sixty-two, which of these sums would you use?" A student suggests the sum of sixty-seven, and Todd says, "It is the only one in the sixties." Then a student suggests that fifty-eight might be closer. The teacher tells them to talk with a partner to decide which sum they think is correct. After discussion, they decide that fifty-eight is closer. Manuel's proof involves counting the spaces between the two sums and 62 on the hundreds chart.

The teacher then explains to students that they will play the game *On Target*. As they play, they will be thinking about different ways to arrange numbers, as they did in this activity. She tells the students, "In the game, though, you will be trying to get a specific sum." After she reviews the directions for the game and distributes materials, play begins.

Tips from the Classroom

- Working in teams gives opportunities for students to discuss their thinking. In the field testing, students often focused on place value, deciding which numbers should go in the hundreds place first, then the tens, and then the ones. West told his partner, "We need to find two numbers that add together to equal five. Since our target number is 512, if we put them in the hundreds place, we should win."
- We found it interesting to note students' reactions to getting a sum greater than the target number. Among themselves, different groups of players made the following decisions: (1) If your sum is greater than the target, you lose; (2) Use a number line, look at the differences, and decide which is less; and (3) If your sum is greater than the target, rearrange the numbers so it is less. We decided not to list a rule about this situation, but rather to allow students to help create the game and to perhaps play differently from other groups in the room.
- In the field testing, students compared numbers to decide which sum was closer, often referring to place value. You may wish to have students add up or subtract to find exact differences.
- If you want to review the recording sheets later, to make sure students added and identified the closer totals correctly, have opposing teams staple their sheets together and give the sheets to you.

What to Look For

- Do players recognize the greater significance of the hundreds digits as compared with the tens or ones, when trying to get close to the target sum?
- Do players notice that exchanging digits in the same place of each addend does not change the sum?
- What strategies do players use to find the sums?
- What do players do when a sum is greater than the target sum?

Variations

- Have students work with two-digit numbers. The group rolls two dice to represent tens and ones, and each team gets four cards.
- Allow each team to arrange the numbers shown on the dice to form the target sum *after* the players have seen the six digits they can use. Note that in this version, the teams might be trying for different target sums.
- Have students use the digits to form addends with one-, two-, or three-digit numbers, whichever produces the closest total.

Exit Question Choices

> The target number is 630, and you are dealt the cards 7, 2, 5, 3, 4, and 2. What addition problem will you make?
> How did place value help you to make decisions during the game?

Extension

Present the following task.

Use the digits 1, 2, 3, 4, 5, and 6.
Write two numbers with three digits each. Do not use a digit more than once.
Which two numbers give you the greatest total? Which two numbers give you the least total?

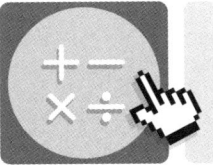

Online Games and Apps

While there are numerous online games and mobile apps that give students the opportunity to practice single-digit addition facts, there are some games that also provide students with visual models, audio support, and opportunities for critical thinking. These online games provide flexibility in such ways as demonstrating the number of objects being added rather than just the digits or showing the addition on a number line or with a ten-frame. Players may often choose a range of numbers for the addends or the sums, allowing for greater differentiation within one game.

- Breakapart, a free game found on the Greg Tang Math website at http://gregtangmath.com/breakapart, engages students in solving addition equations using the break-apart strategy. Players may choose the type of problem they wish to solve, including doubles, make 10, make 100, and partial sums. Essentially playing against the clock, players must choose what number belongs in the box by breaking apart one of the numbers in the expression and then using the new numbers to find the solution to the equation. For example, if the equation is $39 + 7 = $ _____, players first must choose how to break the 7 into smaller numbers in order to efficiently find the total. Choosing $1 + 6$ then allows the players to see $39 + 1 + 6 = $ _____ and to recognize that

39 + 1 = 40, and 40 + 6 = 46. Thus, the correct answer is 46. Incorrect answers count against the player in the form of penalty time. A Hint button is available for those students who need extra assistance.

- Number Bonds II, a free game found in the "Addition and Subtraction Games" section of the Math Playground website (at http://www.mathplayground.com/number_bonds_II.html), provides players with the opportunity to practice finding sums while exploring alternative choices. The game player chooses a target number from ten to twenty. Once the player chooses, the game board is revealed, consisting of single- or two-digit numbers, depending upon the target number chosen. The player chooses pairs of numbers that sum to the target number. For example, if the chosen target number is fifteen, the student must find pairs of addends on the game board whose sum is fifteen. Only pairs of numbers that can be connected by a path consisting of two or fewer physical turns on the board are acceptable. Players make strategic decisions as they look for numbers that have the target sum and also fit the path requirement.

- Fruit Splat Number Line Addition, a free game found on the Sheppard Software website at http://www.sheppardsoftware.com/mathgames/earlymath/fruit_shoot_NumberLine.htm, presents players with the opportunity to practice matching addition equations with their number line representations. The game supports differentiation by allowing players to choose equations with a second addend that is always one, a second addend that is always two, or any combination of addends with a sum equal to or less than ten. Players investigate a number line representation at the bottom of the screen while four or five fruits with addition equations float above the line. Players click on the fruit with the equation that matches the number line. Each correct answer records a hit, while incorrect answers record misses, and a score is tallied. Players may choose to have the fruits float more quickly or slowly and to have the game timed or not.

CHAPTER 6

Subtraction

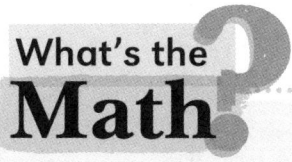

Subtraction is generally introduced after addition and initially modeled as a take-away situation. Students begin with a total set of counters, take away some of them, and find the number of counters that are left. Over time, students also explore subtraction as a missing addend or as a comparison, and counting strategies may replace modeling with drawings or manipulatives. Understanding part-whole relationships helps students develop stronger connections between addition and subtraction.

When using counting, students may count on or count back (Sarama and Clements 2009) to find a difference. To find 7 − 3, a student might count back, saying, "Six, five, four, three," while keeping track that she has said four numbers. Counting forward, a student would count on from three, saying, "Four, five, six, seven," while also keeping track of how many numbers he was saying. Many students seem to rely on the counting-back strategy, which may limit their ability to think flexibly (Rickard 2013). We believe that opportunities to talk with partners in game situations can stimulate the use of alternative approaches.

In kindergarten students generally explore subtraction with totals to 10 and develop fluency with facts to 5. In first grade, students usually explore subtraction with totals to 100 and develop fluency with facts to 20. Students are likely to develop fluency with totals to 100 and extend the exploration of subtraction with totals to 1,000 in second grade. Games and puzzles in this chapter provide students with opportunities to model subtraction, develop strategies for finding differences, recognize related facts, estimate differences, and develop fluency.

How Many Are in the Cup?

Why This Game or Puzzle?

We include this classic game as we believe it provides an excellent opportunity for students to explore subtraction strategies. The game presents a take-away subtraction situation and yet lends itself to thinking about missing addends. Players put five or ten counters in an opaque cup. Team 1 takes some counters out of the cup and places them so that they can be seen by all players. Once the teams agree on how many counters have been taken from the cup, Team 2 must determine the number of counters that are left inside. Together, players may use counting strategies, known facts, or fact recall to identify the number of counters left in the cup. Team 1 counts to check and then Team 2 writes a number sentence to represent the turn. The game ends after players have recorded six number sentences. There is no winner or loser in this game.

"When students are forced to work with greater numbers before they can work fluently with numbers to ten, they can become dependent on rules and procedures that have no meaning" (Postewait, Adam, and Shih 2003, 354). Facts to ten are built on recall of facts to five. Playing *How Many Are in the Cup?* gives students the opportunity to model subtraction and record associated subtraction number sentences as they gain fact strategies with totals to five or ten.

Math Focus

› Modeling subtraction
› Finding differences with totals to five or ten
› Recording subtraction sentences

Materials Needed

› 5 or 10 counters per group
› 1 opaque cup per group
› 1 *How Many Are in the Cup?* Recording Sheet per group (page A-48)
› Optional: 5 or 10 counters per team for modeling
› Optional: 1 *How Many Are in the Cup?* Directions per group (page A-49)

> **Directions**
> Goal: Find the number of missing counters.
> - Put the counters in the cup (either five or ten counters).
> - Decide which team is Team 1. The other team is Team 2.
> - Team 1 takes some of the counters out of the cup and puts them between the two teams. The teams agree how many counters they can all see.
> - Team 2 decides how many counters are still in the cup.
> - Team 1 counts the counters in the cup.
> - Team 2 writes the subtraction number sentence on the recording sheet.
> - Take turns being Team 1 and Team 2.
> - The game ends when, together, the teams have written six number sentences.

How It Looks in the Classroom

A kindergarten teacher invites the students to the rug area and introduces the game using five counters. She puts the counters in a row and invites Jackson to count them, pointing to each one as he counts it. She then invites Jodi to count them as well. Then the teacher puts them into the cup, counting as she does so. She asks, "How many counters are there in the cup?" The students agree that there are five.

The teacher asks the students to watch closely as Benita and Jason play the game with Rosa and Marcus. She tells Benita and Jason to take some counters out of the cup and place them for all to see as they count to tell how many were taken. Together, Benita and Jason take three counters out and count, "One, two, three," as they do so. The teacher then tells Rosa and Marcus that it is their job to decide how many counters are still in the cup and that the other students in the class should turn to their neighbors and make their decisions, too.

After the allotted time, Rosa and Marcus say that they think there are two counters in the cup. When the teacher asks them to tell how they know, Rosa puts out her right hand with her fingers spread out as Marcus says, "This is how many counters there are."

Rosa folds three fingers into her palm while saying, "These were taken out of the cup. Two are left."

The teacher turns to the class and asks for other ways to find the answer. Kane says he drew five counters and crossed out three of them to see the two that were left. Jassie says she just knew the answer. Michael says, "I counted, 'Four, five,' and knew there were two counters left."

Then the teacher asks Benita and Jason to empty the cup, putting the counters from the cup in a separate pile. "How many do you see?" she asks. All agree that two counters had been left in the cup.

The teacher records the number sentence 5 − 3 = 2 on nearby chart paper and asks, "How does this equation relate to the counters and the cup?" Once she is sure that the students associate the 5 with the total number of counters, the 3 with the number of counters removed from the cup, and the 2 with the number of counters that were left in the cup, she chooses other partners to play another round of the game. Then the teacher reviews the directions and distributes materials for teams to play the game on their own.

Tips from the Classroom

- Some students responded to the task by trying to guess the number of counters left in the cup. We found that teammates usually suggested other approaches. We remember Eric, who said, "Wait; let's try to count and find exactly how many are left." If guessing persists, prod students by asking questions such as *What could you do to find the exact number?* and *How can you prove that is the answer?*
- Have sets of five or ten counters available for some players to use as they try to determine how many are left in the cup. Some students may wish to organize the counters on a five- or ten-frame.
- Encourage players to talk about how they are identifying the number of counters in the cup.

What to Look For

- When students are shown sets of counters, do they count correctly?
- What strategies do students use to determine the number of counters in the cup?
- Are students' number sentences correct?
- What can you learn from the students' recording sheets? Figure 6.1 shows a recording sheet from a group that wanted to work together to find all of the ways they could subtract from five.

How Many Are in the Cup? Recording Sheet

Name(s): _____ Date: _____

1) 5   0   5
2) 5   2   3
3) 5   1   4
4) 5   4   1
5) 5   3   2
6) 5   5   0

Figure 6.1 How Many Are in the Cup? recording sheet

Variations

- Have students draw to represent each turn rather than write a number sentence.
- Start with ten counters visible to both teams; have one team hide some of them under the cup while the other team isn't looking; and have the other team determine how many of the counters are hidden.
- For a greater challenge, have students play with twenty counters.

Exit Question Choices

> The picture shows how many of the five (or ten) counters Toni took out of the cup. How many counters are in the cup?

> There are ten (or five) counters in the cup. Team 1 takes four counters out of the cup. Write a subtraction sentence to find how many counters are in the cup.

Extension

Have students respond to word problems such as the following one.

There are 10 bikes for sale at the yard sale. Some are bicycles and some are tricycles. How many of each could there be?

This game is adapted from Kathy Richardson's *Hiding Assessment* (2003).

Why This Game or Puzzle?

Move Along also focuses on subtraction with facts to ten, but, in this game, manipulatives are not required. Baroody (2006) suggests that students' fact knowledge evolves through three stages: use of counting; use of strategies or reasoning; and efficient retrieval. It is more likely that students' fact knowledge will evolve when they have opportunities to discuss and practice subtraction strategies with peers.

In this game teams try to move their game pieces along the path on the game board until they reach the Home space. On each turn, players draw two number cards and subtract each number from ten. If at least one of the differences is listed in the next space, the team may move its game piece forward to that space; otherwise, the team does not get to move its piece.

Math Focus
› Finding differences
› Deciding which number to subtract to get a given difference

Materials Needed
› 2 game pieces per group
› 1 *Move Along* Game Board per group (page A-50)
› 1 deck of *Move Along* Cards per group (page A-51)
› Optional: 1 *Move Along* Directions per group (page A-52)

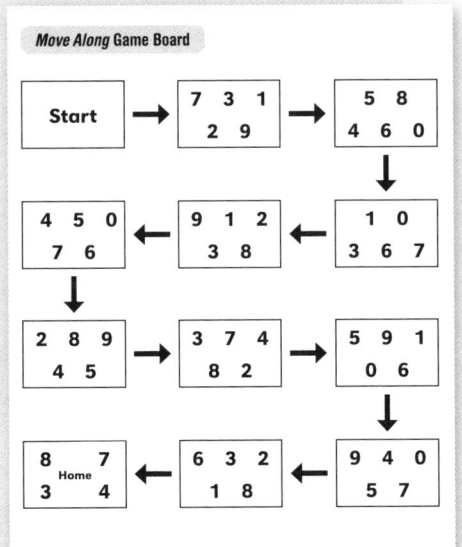

Directions
Goal: Be the first team to land on the Home space.
› Each team puts its game piece on Start.
› Mix up the game cards and spread them out facedown between the teams.
› Take turns.
› On each turn:
  › Pick up two cards. Choose a number on one of the cards to subtract from ten.
  › If the answer is in the next space on the game board, move your team's piece to that space.
  › If the answer is not there, subtract the other number from ten.
  › If that answer is there, move your team's game piece to that space.
  › If neither answer is there, your team's game piece stays where it is, and your turn is over.
› After each turn, put your cards facedown and mix up all the cards.
› Whichever team makes it to the Home space first wins the game.

How It Looks in the Classroom
A first-grade teacher writes the numbers *5, 3, 7, 1,* and *4* on a piece of paper placed under the classroom projection device. She shuffles the *Move Along* cards, draws one of them, and places the card faceup under the projection device. She asks, "If we subtract this six from ten, is the answer one of these numbers? If so, which one?"

After a brief time the teacher asks students who found an answer to write it on their whiteboards and hold them up so she can see them. As the teacher glances at the responses, she notes that nearly all of the students have written *4*, while one student did not record anything, and two students wrote *3*. The teacher holds up a filled ten-frame and asks, "Who can use this ten-frame to prove your answer?"

Jasmina says, "We can see the ten. I can split the six to subtract it. I think of the top five being gone and one more in the next row. That leaves four."

"Zuri, can you explain Jasmina's thinking in your own words?" the teacher asks.

Zuri says, "You can think of six as a five and a one. If you take away the top row, you need to take away one more. That leaves four."

The teacher then writes five new numbers less than ten, turns over another card, and repeats the activity. Satisfied that the students understand the process with one card, the teacher displays the *Move Along* game board and explains that in this game, they'll turn over two numbers. She tells them that their task will be to try to find at least one of these numbers that they can subtract from ten and see the answer in the next space. The class plays a few rounds of the game as two large teams and then the students play in teams of two.

Tips from the Classroom

> Some students will be more successful if counters and ten-frames are available for use.
> There are times in the game when one team's piece can obscure numbers on the space to which the other team hopes to move. Have teams use transparent colored discs or have players move their piece just above the space, so the other team can see the numbers.
> Encourage partners to identify the numbers they hope to get on their turns.
> If you plan to have students respond to the second suggested exit question, tell them the question before the game begins so that they can think about their fact knowledge as they play. You may wish to make sticky notes available for students so that they can note examples of facts they know or have to figure out as they play.

What to Look For

> Do partners randomly choose a number to subtract or recognize a choice that will allow them to move their game piece forward?
> What subtraction strategies do students use?
> What facts can students recall?
> Do students subtract accurately?

Variations

> Change the game board so that each space lists only two numbers from 0 to 5, and have players subtract from five.
> For students ready for a greater challenge, change the numbers on the game board and the cards to be multiples of ten, and have the players subtract from one hundred.

Exit Question Choices

> You are one space before the Home space. What number would you like on one of your cards? Why?

> What facts did you know without having to think about them? What facts did you have to figure out?

Extension

Have players create fact recall goals and write them in their journals or tell them to you to record. Encourage them to also discuss how they will reach their goals.

Take the Numbers

Why This Game or Puzzle?

There are a variety of ways to think about subtraction. A student might find 6 − 2 by taking two counters away from six counters and finding the number of counters left or by counting backward to identify two less than six. Alternatively, a student might show two counters, add counters until there is a total of six, and then identify how many were added, or count forward from two up to six and identify how many numbers she said. Torbeyns and colleagues (2009) refer to the taking-away situation as direct subtraction and the missing-addend approach as indirect subtraction and suggest that direct is easier than indirect. Either subtraction strategy may be applied in this game, as well as mental arithmetic or paper-and-pencil strategies.

In this game a set of cards labeled *11–19* are referred to as the board numbers. Each team gets four cards from a deck of playing cards. (Face cards have been removed; aces stand for 1.) When a team places the first playing card beneath a board number, it finds the difference of these two numbers. The team then finds the board number on the *Take the Numbers* recording sheet, crosses it out, and records this difference below it. (The recording sheet provides space for two games.)

Teams take turns placing a playing card beneath a board number and finding the difference. When a team places a playing card below a board number that already has one or more playing cards placed beneath it, the team subtracts the number on its card from the last recorded difference in that board number's column. When a team places a card that results in a difference of zero, that team turns over that board number and collects all the playing cards for that number.

Because players can see each team's cards, they may make strategic decisions about which of their playing cards to place under which board numbers.

Math Focus
› Finding differences from totals to nineteen

Materials Needed
› 1 set of *Take the Numbers* Board Number Cards per group (page A-53)
› 1 deck of playing cards, without the face cards (aces stand for 1), per group
› 1 *Take the Numbers* Recording Sheet per group (page A-54)
› Optional: 1 *Take the Numbers* Directions per group (page A-55)

Directions
Goal: Collect the greater number of playing cards.
› Put the nine board numbers faceup in a row so everyone can see them.
› Mix up the playing cards. Put them facedown in a deck.
› Deal four playing cards to each team faceup for all to see.
› Take turns. On each turn you place one of your playing cards below one of the board numbers. You can put your playing card below a board number that already has a playing card there or below one that does not. Then look at the recording sheet for that number.
› If you put the first card below a number, subtract the value of your playing card from the board number you chose. Cross out the board number on the recording sheet and write the difference you found in that column. Then take a new playing card from the deck to add to your hand.

> - When there is already a card below the board number you chose, check the recording sheet. If your card is less than the last number in the column on the recording sheet for the board number you chose, subtract your card from that number. Cross out the last number in that column and record the new difference.
> - If your card is equal to the last number in the column, subtract and record the zero difference. Turn over that game board number, as it is no longer in play, and take all of its playing cards. Put them facedown near your team. Take a card from the deck to put in your hand. Your turn is over.
> - If you can't place a number that gives a difference greater than or equal to zero, you lose your turn.
> - The game ends when neither team can place a card. The team that has collected the greater number of playing cards wins.

How It Looks in the Classroom

When math time begins in this first-grade classroom, the teacher displays the following task, which she reads aloud to the students. Then the students and the teacher read it together once more.

Write about what you know about subtraction and zero.
What happens when you subtract zero from a number?
When do you get a difference of zero?

Before discussing their responses in the larger group, the teacher asks them to share their ideas with their neighbors. As she walks around, she overhears statements that let her know that many of the students are making generalizations such as "Taking nothing doesn't change anything," and "You need to take it all to get left with nothing." When the students share their ideas about each question in the larger group, she records their statements on chart paper. She wants her students to be exposed to many ways of expressing their mathematical thinking, so once she has recorded all responses to a question, she reads them back to the students.

Next the teacher records the open number sentences shown in Figure 6.2. She points to the first one and asks, "What numbers would make this sentence true?"

Several students give specific examples, such as "They can both be fours," and then Elijah says, "It can be anything as long as they are both the same."

The teacher then points to the second example and again asks for numbers that would make the sentence true. August suggests that two sixes would work, and then Kaylee exclaims, "This is like the other one. It will always work if both numbers are the same."

The teacher introduces *Take the Numbers*, satisfied that they will play the game with greater awareness about zero and subtraction.

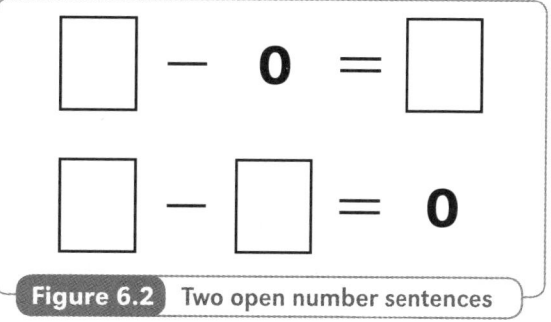

Figure 6.2 Two open number sentences

Tips from the Classroom

- Encourage both teams to check that the differences are accurate.
- Initially a team may decide which of its cards to play without considering the other team's cards. We found that players learned the importance of thinking about those cards when the other team was readily able to play a card that resulted in a difference of zero.

What to Look For

- Do students easily recognize when they have a number that will result in a difference of zero?
- What conversations do you hear between team players that you would like students to share with the larger group?
- What types of strategic decisions do players make? Do they consider the other team's cards before placing one of theirs?

Variations

- Change the rules so that if a team gets a difference of zero, the other team gets those playing cards.
- The students in one class suggested that teams place their initials next to a zero difference when their card resulted in that difference. Then the team with the greater number of initialed zeros would win.
- Have students play without using the recording sheet to see how they determine the difference that remains when more than one card has been placed beneath a number on the game board.

Exit Question Choices

- A 3 and a 6 are under the 14. What number do you need to get a difference of zero? How do you know?
- There are two numbers placed under the 16 on the game board. Then you place a 1 there to make a difference of zero. What two cards could have been there before you placed the 1? What would be another answer?

Extension

Have students talk or write about the following:

Your partner doesn't understand why you should look at the numbers on the other team's cards before deciding which card you should place on the board. What would you say to help your partner understand why this is important?

Name That Number

Why This Game or Puzzle?

Communicating mathematical ideas and sharing mathematical thinking are essential habits of mind. For this puzzle, team members must share what they know in order to have all of the necessary data to find the solution. The need to collect data is relevant to almost all real-world problem-solving situations, as is the need to cooperate; they are critical skills for mathematicians of all ages.

Here, a logic number puzzle is solved cooperatively, based on a model for cooperative problem solving suggested by Tim Erickson (1989). Each member of a three-person team receives two clues about a number. The players must decide how to share and organize the information as they try to identify the number that meets all of the clues. Three levels of the puzzle are provided (pages A-56–A-58). It is likely that the first puzzle is most appropriate for kindergartners, the second one for first graders, and the third one for second graders, though you may find at least two of the puzzles at the right level for some of your students.

Name That Number Clues A

The number is greater than 9 − 5.	The number is less than 10 − 1.
The number is not 10 − 4.	The number is not 8 − 0.
The number is not 9 − 2.	What is the number?

Math Focus

- Subtracting one-, two-, or three-digit numbers
- Estimating differences

Materials Needed
- 1 set of *Name That Number* Clues (A, B, or C) per group (pages A-56, A-57, or A-58)
- Optional: 1 *Name That Number* Directions per group (page A-59)

Directions
Goal: Use the clues to find the mystery number.
- Work as a team of three puzzle solvers.
- Place the clues facedown. Each solver randomly takes two of the clues.
- Decide how to share the clues.
- Work together, read the clues as many times as necessary, and talk about what you know. Try to find the number that fits all the clues.
- When you think you have the solution, read the clues again to check.

How It Looks in the Classroom
One second-grade teacher offers students this logic puzzle to solve:

> *It is greater than 430 − 200.*
> *It is not 339 − 108.*
> *It is less than 689 − 456.*
> *What is the number?*

He has purposely kept the subtraction relatively simple because he wants to focus on the language within the clues and strategies for solving such puzzles. He directs the students to work independently.

After giving students time to solve the puzzle, he asks students for their answers. The students give three different answers: 230, 232, and 233; 230 and 232; and 232. The teacher records the possible answers as they are given and asks for students to raise their hands if they agree. At least three students choose each response.

The teacher asks, "What did you learn from the first clue?"

Students correctly identify that the number is greater than 230. When the teacher asks how he might represent that information, Sam suggests writing *>230* next to the clue. Through discussion, the class decides that writing *231* and then crossing it out would be best for the second clue and that *<233* would summarize the third clue. The teacher recognizes that some students have included 230 and/or 233 in their answers. He knows that identifying which of two numbers is greater is less challenging than recognizing all the numbers indicated by "greater than" or "less than" clues. To clarify the students' thinking, he says, "Tell me some numbers greater than 230." Students offer a variety of responses, none of which are 230. He then asks students to consider their original answers.

Richard says, "I think I want to change my mind. I thought it could be 230 and 232, but 230 is not greater than 230."

The teacher asks, "Does anyone else want to change their mind?" Before long, the class decides that the number is 232. The teacher tells the students that they will now solve another logic puzzle. He says, "This time you will work in groups of three. You will each get two clues. You will decide how to share and organize the clues to find the answer. Take notes so that I can follow your thinking. Happy puzzling!"

Tips from the Classroom

- Some students are likely to want paper and pencil for finding differences or for listing possible numbers and then crossing off candidates as they are eliminated through the clues. In our field testing, one class decided to designate the puzzler who got the *What is the number?* clue as the recorder.
- Some students may wish to have counters or base ten materials available.
- Sometimes one of the puzzlers recognizes the answer before the others. Make sure each player can explain why that answer is correct.

What to Look For

- Do students recognize that "greater than" and "less than" relationships do not include the number specified? Is there other language in the clues that students find challenging?
- How do puzzlers organize the clues? Do they just leave them in the order they are read or categorize them strategically?
- Do students rely on a drawing, model, diagram, written computation, mental computation, or recognition to find a difference?
- How do students work together? Do they make sure everyone participates? Do roles, such as facilitator and note taker, evolve as the students work on the puzzle?

Variations

- Have solvers show their clues one at a time, without talking or gesturing. After all the clues have been shown, each solver may rearrange the order of one of the clues. Through eye contact, rather than talking, solvers communicate that they think they have a solution. Solvers may then talk to check the clues.
- If you choose to make your own *Name That Number* cards, you could create examples that would result in two to four numbers that met all criteria and change the question to *What are the numbers?*

» Have groups record their thinking of how they used their clues to figure out the puzzle. Figure 6.3 shows the thinking of one group. Note the use of number lines to record their steps.

Exit Question Choices
» How might you represent the information you gain from the clues *It is greater than 10 – 7* and *It is less than 10 – 1*?
» How did your team decide to organize your clues?
» What was helpful about working together to solve the puzzle? What was challenging?

Extension
Have students contribute a phrase or a picture to a poster titled "Cooperative Puzzle Solving."

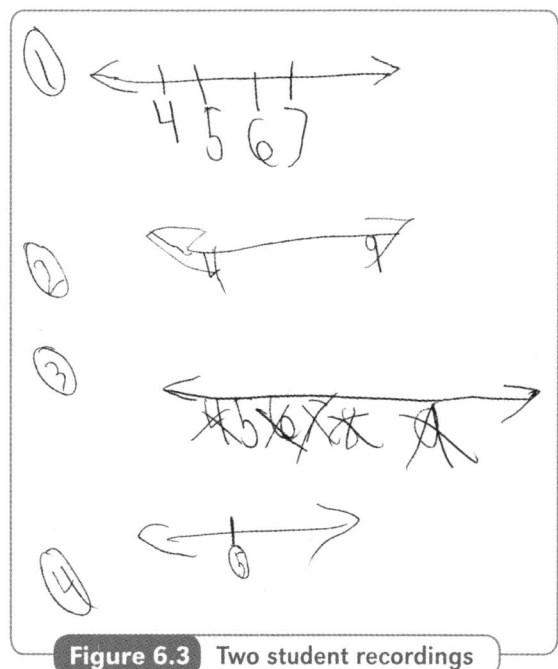

Figure 6.3 Two student recordings

Subtraction Tic-Tac-Toe

Why This Game or Puzzle?
Estimation skills are essential; we typically use them every day. We estimate when exact answers are not required and to check the reasonableness of computation completed with paper and pencil or an electronic device. For instance, in a grocery store, we may estimate to see if we have enough money for what we are purchasing and then estimate again to make sure the total the cashier tells us makes sense. According to Lan and colleagues, "Computational estimation is a complex skill involving many of the same subtleties and complexities as problem solving" (2010, 1). Unfortunately, most of the opportunities students have to practice their computational skills are focused on finding exact answers.

In this game, students are given two sets of numbers. They choose a number from each set, find the difference between them, and write an *X* (or an *O*) on that difference on the answer board. The goal is to write an *X* (or an *O*) on three adjacent differences in a row, column, or diagonal. This aspect of the game encourages students to estimate differences when choosing numbers, make conjectures, plan ahead, and decide when they want to block an opponent's move. When we field-tested the game, students recommended we make the game board look more like a traditional tic-tac-toe game, so we eliminated the outside borders. Three versions of the game are included, offering different levels of challenge.

Math Focus
> Estimating differences
> Finding differences

Materials Needed
> 1 *Subtraction Tic-Tac-Toe* Game Board (A, B, or C) per group (page A-60, A-61, or A-62)
> Optional: 1 calculator per team
> Optional: 1 *Subtraction Tic-Tac-Toe* Directions per group (page A-63)

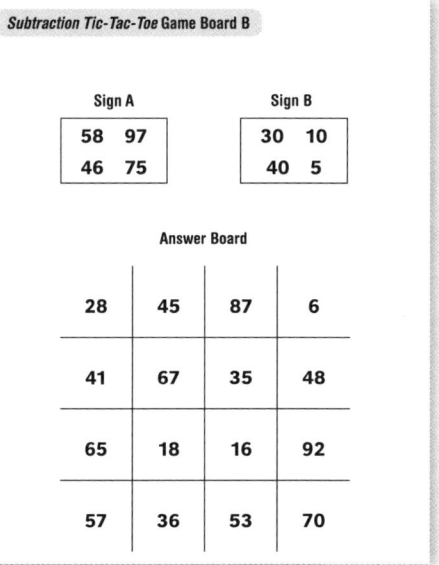

Directions
Goal: Choose pairs of numbers to subtract to mark three differences in a row, column, or diagonal on the answer board.
> Decide which team will be *X* and which will be *O*. Take turns.
> On each turn, the team picks a number from Sign A and one from Sign B. Then both teams subtract the number on Sign B from the number on Sign A.
> Once both teams agree on the difference, the team whose turn it is finds it on the answer board and writes its *X* or *O* on the number.
> If the team gets a difference that is already marked with an *X* or *O*, it loses its turn.
> The first team to write *X* or *O* in three touching differences in a row, column, or diagonal is the winner.

> These *X*s are in the same row and touch:

> These *X*s are in the same row but do not all touch:

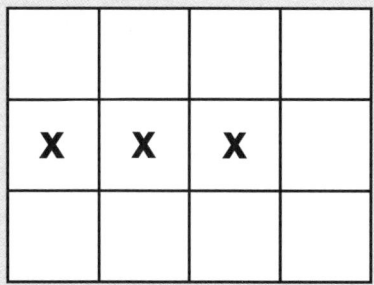
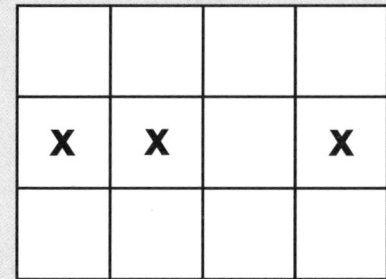

How It Looks in the Classroom

One second-grade teacher displays the expression 763 − 129 as well as the three numbers 646, 892, and 634. She says, "I am going to give you a few seconds to choose the number that you think is the difference."

The teacher counts silently to three and then says, "I want you to put one finger on your chin if you chose the first number, two fingers if you chose the second number, and three fingers if you chose the third number. Get ready. Put your fingers on your chins now."

Once the teacher has this broad perspective of student success, she asks them to explain their thinking. Their responses show a wide range of thinking. Wes reports that he didn't have enough time to subtract, so he made a guess. Lucia explains that she didn't have time to subtract either; she chose 646 because she knew it had to be in the 600s. Cam says, "I choose 892 because 7 and 1 equals 8. Oh no—I was adding and this is about subtraction."

Gracie explains, "I chose 634 because I was thinking, 'What plus 9 ends in 3?' Once I knew it had to be 13, I knew there had to be a 4 in the ones place."

The teacher says, "You all used your number sense to try to predict the difference. We are going to play a tic-tac-toe game where you will also predict differences." She displays the game board (page A-62) and reviews the rules of play, making sure students understand what it means for differences to be "touching." After a few rounds of play as a large group, players form teams of two. Each group gets a copy of the game board and begins playing.

Tips from the Classroom

- Some students will be more successful if hundreds charts and base ten materials are available for use.
- Some students are likely to begin by choosing numbers randomly. It's best not to intervene too quickly.
- It can be helpful to ask players to think aloud as they choose their numbers, as their thinking will give other students access to a variety of strategies.
- You may need to remind players that both teams are expected to find each difference, as a way to check for accuracy. Sometimes opponents become so focused on their next move that they forget this aspect of the game.
- You may wish to allow some or all teams to use calculators to check differences.

What to Look For

- Do students choose numbers randomly or do they use strategies for making choices?
- What evidence of good number sense do you observe?
- What mathematical language do students use to describe their thinking?
- What strategies do students use to help them get three touching numbers in a row, column, or diagonal?
- Do students remember to block opponents when necessary?
- What do teams do if they disagree about a difference?

Variations

- In our field testing, one class suggested this game rule: If a team finds an incorrect difference, it loses its turn. This rule motivated opposing teams to check calculations.
- For game boards B or C, you can add more challenge by changing the goal to marking four differences in a row, column, or diagonal.
- By changing the numbers on the signs and the game board, you can vary the difficulty level of the game.
- You can also create a version of the game in which the players choose addends on signs to get sums on the answer board.

Exit Question Choices

- What strategies did you use to choose the numbers on your turns?
- When did you think it was important to block an opponent rather than find a difference that touched the differences you already had?
- What did you do if you and your opponents disagreed on a difference?

Extension

Have students discuss or record a response to this question:

> *Some players always want to block the other team. What would you tell them about why this may not always be a good strategy?*

This game is adapted from *Four in a Row* in *Zeroing in on Number and Operations: Key Ideas and Common Misconceptions, Grades 3–4* (Dacey and Collins 2010b).

Online Games and Apps

Continual practice with whole number operations in a variety of contexts allows young students to become more comfortable with the structure of each operation. Whenever possible, it is desirable to choose online games that allow for different visual representations, a variety of levels, and various problem types, thereby utilizing the power of the technology. Following are some examples:

- How Many Under the Shell? is a free game found on the National Council of Teachers of Mathematics' Illuminations website (http://illuminations.nctm.org/Activity.aspx?id=3566) that is similar to *How Many Are in the Cup?* This game is beneficial when visual models are more appropriate for students than concrete ones. It engages students with Okta the octopus and his subtraction problems, which are visually represented by bubbles hidden under a shell. Game partners may choose the number of bubbles that Okta starts with or let the game begin with a random number. Okta shows some bubbles, hides them under the shell, and then shows how many he takes away. Each bubble initially includes a number to show the number of bubbles being counted as they move under the shell. Players must complete the resulting subtraction problem displayed by Okta's lobster friend. Incorrect answers are responded to only with a sad-faced Okta, whereas correct answers show all bubbles revealed by a happy-faced Okta.

- Diffy, a free virtual puzzle from the National Library of Virtual Manipulatives at http://nlvm.usu.edu/en/nav/frames_asid_326_g_1_t_1.html?from=category_g_1_t_1.html, starts by showing four squares nested inside each other. The outer square shows a number at each vertex. Players determine the difference between each set of two adjacent numbers. When players type in the correct differences, these differences are shown in the center of each side of the square, which is also in the vertex of the next square inside. This process continues until the differences within four squares are completed. Each puzzle is different and gives players the opportunity to observe patterns from one puzzle to the next. You may wish to encourage players to predict what the final difference will be prior to completing a puzzle. As an additional variation, students may create their own puzzles by choosing numbers for each of the corners of the beginning square.

- Speed Grid Challenge: Subtraction (1), a free game found on the Oswego website at http://www.oswego.org/ocsd-web/games/speedgrid/subtraction/urikasub1res.html, gives players practice with choosing two of the three numbers in a subtraction problem from a grid of numbers shown on the screen. For example, players may see _____ − _____ = 8 at the bottom of the screen and have to choose from a grid of sixteen numbers in order to fill in the blanks. There are a variety of possible responses. The game allows students to choose the number of problems they must answer and the time frame in which they must answer the problems, allowing for differentiation among players. Encourage players to discuss their strategies for choosing numbers.

CHAPTER 7

Addition and Subtraction

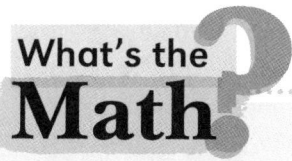

Sharon Griffin (2004) suggests that teaching with a focus on conceptual relationships, rather than procedural rules, supports students' number sense. Games and puzzles involving addition and subtraction provide opportunities for students to further investigate relationships between these operations, while strengthening their conceptual understanding. Through experimentation and discussion, for example, students can develop the ability to understand that if $4 + 6 - 2 = 8$, then $4 + 6 - 3 = 7$, without actually computing. They begin to recognize that by subtracting three instead of two, the resulting value decreases by one. Students can also gain operational sense. If the numbers 2, 7, and 6 are to be used in an expression equal to 11, students should recognize that more than one operation would be needed. Addition must be involved in some way because eleven is greater than any one of the whole numbers. Subtraction must also be involved, as eleven is less than the total of the three whole numbers.

With the activities in this chapter, students solve word problems, identify fact families, develop and apply number sense, and think more deeply about the relationships within and between addition and subtraction. There is also an emphasis on deductive reasoning, conjecturing, and proving or disproving. Hillen and Watanabe (2013) identify such thinking as the essence of mathematics.

Word Problem Bingo

Why This Game or Puzzle?

In many classrooms a single word problem might be the focus of a mathematics lesson. Students work together to interpret the meaning of the problem and share a variety of strategies for solving it. Traditionally, addition or subtraction word problems, no matter the problem type, ask only about the result (Van de Walle, Karp, and Bay-Williams 2013). It is important that students explore problems in which any one of the three possible numbers (start, change, and result or part, part, and whole) is missing. Students also need ample opportunities to solve a variety of problems, not just when being introduced to a new problem type or computation strategy.

In this game students write *FREE* in one of the empty spaces on the bingo game board and then randomly write the numbers *1–15* in the remaining spaces. The teacher, another adult, a reading buddy, or a player-leader (a student who both plays on a team and leads) reads a story problem, and teams work together to solve it. When they do, they mark the number representing the numerical answer to the word problem on their boards. The first team to mark four numbers correctly (or three problems and a free space) in a row, column, or diagonal on its game board wins.

Math Focus

- Solving addition and subtraction word problems
- Explaining problem solution strategies

Materials Needed

- 1 *Word Problem Bingo* Game Board per team (page A-66)
- 1 deck of *Word Problem Bingo* Cards per group (pages A-64–A-65)
- Optional: 1 *Word Problem Bingo* Directions per group (page A-67)

Directions

Goal: Be the first team to get four answers (or three answers and a free space) in a row, column, or diagonal on your game board.

- Choose a player-leader from one of the teams to both play and lead.
- Write *FREE* in one of the spaces on your team's game board.
- Write the numbers *1–15* randomly in the spaces that are left. Do not write the numbers in order.
- Mix up the cards and place them facedown.
- The player-leader turns over a card, reads the problem, and leaves it faceup.
- Both teams solve the problem and talk about the answer.
- When all players agree, the player-leader records the numerical answer on scrap paper, and each team puts an *X* on the answer on its game board.
- The first team to get four *X*s, or three *X*s and a free space, in a row, column, or diagonal says, "Bingo!"
- Together the teams check the bingo numbers with those the player-leader recorded. If both teams agree that the answers are correct, the team that said, "Bingo," wins.

How It Looks in the Classroom

A second-grade teacher knows that students often find story problems more challenging when the first number in a problem is missing. She wants to help ensure that students think about relationships among the three numbers in a one-step addition or subtraction story problem before solving different types of problems independently. She writes the following on the board:

_____ people are waiting at a bus stop. _____ of them get on the next bus that stops. _____ people are still waiting for a bus.

She asks for volunteers to explain the bus stop situation using their own words. After two students respond, Trayvon summarizes by saying, "There are some people at a bus stop. Some of them get on the next bus. Some of them are still there waiting."

The teacher then writes *4* in the second blank and *3* in the third blank and asks, "What number should we write in the first blank?"

The students turn to their neighbors and solve the problem. Once the answer is identified as seven people, they share their solution strategies. Different techniques are discussed, including making a drawing, counting up three from the four, and finding four plus three.

The teacher gives each set of partners a copy of the game board and tells them to write *FREE* in one of the spaces and randomly fill in the numbers *1–15* in the other spaces. She reads one of the word problem cards and then places it under the document camera for all to see. After students have solved it, she has a volunteer explain his or her solution, proving that it is correct. Once everyone agrees, she tells the students to mark the answer on their game boards, and she records the number on the whiteboard. This process continues until one of the teams proclaims, "Bingo!" That team reads its answers, and when the class confirms the numbers are included in those recorded on the whiteboard, the team's win becomes official. The teacher decides to play again as a whole class tomorrow and then, when the fifth-grade reading buddies arrive tomorrow afternoon, she will have them support play in small groups with partners working together to solve the problems.

Tips from the Classroom

- Make sure that teams talk about their answers before marking numbers on their game boards.
- We found that many students preferred to draw a symbol, such as a heart or smiley face, rather than write *FREE*.
- Some students can find it challenging to keep track of the numbers 1–15, when writing them randomly. You can suggest that they write the numbers *1–15* at the top of the game board page and cross them out as they write them in the game board spaces.
- Students are more successful and engaged when problem contexts are familiar and names in problems correspond to those of classmates. Feel free to change the problem contexts or names to better capture the interests and experiences of your students.
- You may wish to continue whole-class games, or have only one group play at a time, so that you or a player can put each problem under the document camera for all to see easily.
- You may wish to have a few parent volunteers or upper-elementary school "buddies" join your class and serve as problem readers.
- Students can use the same cards for several games. If you want students to play more games, replace the problems with ones you provide.

What to Look For

- What strategies do students use to solve the problems?
- Are there certain problem types that are more or less challenging for students?
- How do students prove or disprove that their answers are correct?

Variations

- You can change the numbers in the problems and the answers to be recorded.
- For extra challenge, include some problems that require both addition and subtraction for their solution or problems that have three addends.

Exit Question Choices

> Your team is solving this problem:

Cam bought 5 small plants and some large ones. He bought 7 plants in all. How many large plants did Cam buy?

What are two different equations you could write to solve this problem?

> Choose one of the problems you solved in the game. Write an explanation of how you solved it.

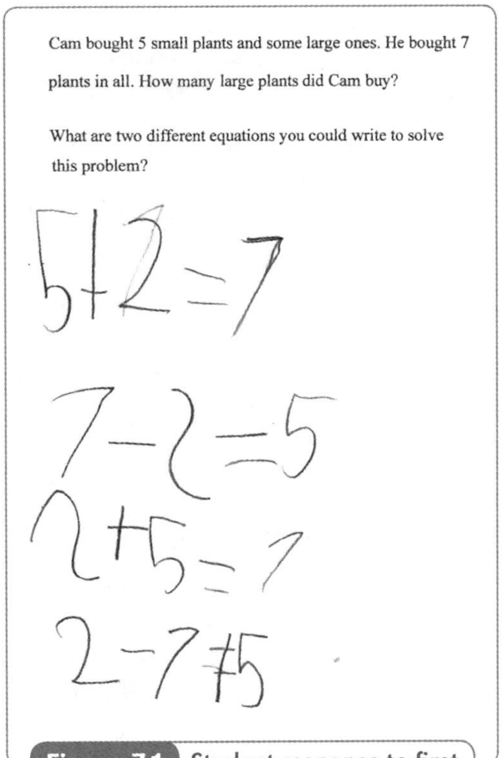

Figure 7.1 shows an example of a student response to the first exit question that illustrates how these questions can sometimes uncover unexpected knowledge. The student quickly noted that there were more than two equations he could use to solve the problem. He started with $5 + 2 = 7$ and $7 - 2 = 5$. He then said, "I can do more." As he wrote $2 + 5 = 7$, he said, "That is just the opposite."

Then he wrote $2 - 7 = 5$, hesitated, and said, "Hmm, two minus seven does not equal five. So, how can I write that?"

The teacher showed him the symbol $\neq$ and explained that it meant *does not equal*. The student quickly completed the final equation, and the teacher asked, "How did you know they were not equal?"

He responded, "When you take more than you have, it's a negative."

Figure 7.1 Student response to first exit question

Extension

Have students write their own word problems to be included in the next *Word Problem Bingo* game.

This game is adapted from "Subtraction Is More Than Take Away" in *Zeroing in on Number and Operations: Key Ideas and Common Misconceptions, Grades 1–2* (Dacey and Collins 2010a).

Four of a Kind

Why This Game or Puzzle?

Visual number bonds, shown horizontally or vertically, can be used to summarize the addition and subtraction relationships among two addends and their sum. Figure 7.2 shows a possible number bond for 12; it indicates that $4 + 8 = 12$; $8 + 4 = 12$; $12 - 4 = 8$; and $12 - 8 = 4$. Students usually find it less challenging to recognize the equations in a fact family when this type of visual model is available. Without it, they often rely on partial understandings. A student might know that two addition and two subtraction sentences should be identified, for example, but record $15 - 8 = 7$; $7 + 8 = 15$; $9 + 7 = 16$; and $16 - 7 = 9$. "It's important to help students make connections among mathematical ideas so they do not see these ideas as disconnected facts" (Burns 2007, 16).

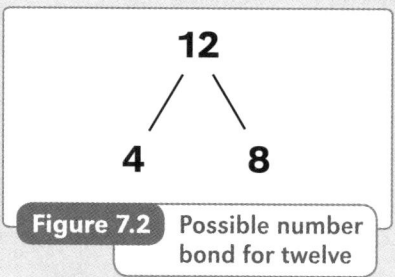

Figure 7.2 Possible number bond for twelve

We've taken a game we learned from Michael Schiro (2009) and adapted it to emphasize the important links between addition and subtraction as well as to provide practice. The goal of the game is to make a set of four equations with the same missing number, all of which are within the same fact family.

Math Focus

› Recognizing fact families
› Identifying missing numbers in addition or subtraction equations

CHAPTER 7
Addition and Subtraction

Materials Needed
› 1 deck of *Four of a Kind* Cards per group (pages A-68–A-69)
› Optional: 1 *Four of a Kind* Directions per group (page A-70)

Directions
Goal: Be the first to get four cards with equations that have the same missing number.
› Four players sit in a circle.
› Shuffle the cards.
› Deal four cards to each player.
› Players look at their own cards and decide on one card they do not want. Each player places this card facedown in front of the player to his or her right. Players put their new cards in their hands.
› Players continue to pass and pick up cards, waiting for all players to pick up before the next pass begins.
› The first player to get four cards with the same missing number says "Four of a kind" and wins.

How It Looks in the Classroom
One first-grade teacher asks her class, "What do you know about fact families?"

Kassi responds, "There are four equations in them."

Caleb says, "Two are addition, and two are subtraction."

Norah says, "I know all the numbers have to be related to each other."

The teacher asks, "Are two, five, and nine related?"

Norah says, "No, they have to equal each other. Like two, five, and seven in a number bond."

The teacher draws this number bond on the board and asks students to write the four equations the bond represents.

The teacher knows that the students need to think about fact families some more and that this game will help them to do so. She organizes them in groups of four and gives each group a deck of *Four of a Kind* cards. To familiarize them with the cards, she tells students to shuffle them and organize them by their missing numbers. Once they've all sorted the cards, she asks them how they found the missing numbers.

Mandi says, "I thought about how the numbers were related."

James says, "I thought about things I know. So for 'blank' minus 7 equals 8, I thought about 8 plus 7."

The teacher then explains the rules of the game, the groups reshuffle their cards, and play begins.

Tips from the Classroom

- Some students were unable to find four of a kind without recording the missing numbers on the cards. You may wish to laminate the cards and allow this for the first few times the students play the game.
- To keep the timing of the passing in sync, you may want to have a player in each game call out "Pass" and "Pick up" as appropriate.
- Some students really like competitive games to involve speed and may rush the other players. If you want such students to slow down, consider introducing the cooperative variation described later in the *Variations* section.
- We found that when some players identified a pair of cards they saw as being of the same kind, they set them aside. Sometimes this was a good choice as it focused their thinking, and sometimes it limited the team as another set might have been easier to make. To stimulate players' reflection on this strategy, ask, *How does putting those cards aside help you? How does putting the cards aside limit your choices?*
- This game is played individually. You may wish to have one or two individuals play in teams, keeping the total number to no more than six players.

What to Look For

- Do students use facts they know to find other missing numbers?
- Do students recognize fact families? Do they draw number bonds to help them recognize equations within a fact family?
- Do players make good choices when they pass cards? That is, do they pass a card without a match or with the same number of matches as other cards they have in their hands?

Variations

- Differentiate, as needed, by creating new cards for the game.
- Have students play cooperatively. Hands of cards stay faceup for all to see throughout the game. There is no talking allowed in the game, but students can gesture to each other. The goal is to have everyone get a set of matched cards in the fewest number of passes.

Exit Card Choices

- How would you rewrite $5 + \_\_\_ = 12$ as a subtraction equation?
- What mathematics might you use to find the missing number in the equation $\_\_\_ - 3 = 9$?

Extension

Have students dramatize a meeting between Ms. Addition and Mr. Subtraction. What might they share about what they have in common? What might they say about how they are different? What math terms might they want to explain as they get to know each other? This game is adapted from *Pass It* in *Mega-fun Math Games and Puzzles for the Elementary Grades* (Schiro 2009).

It's Greater

Why This Game or Puzzle?

Developing students' understanding of addition and subtraction, rather than having them merely apply rote procedures, is a complex goal. To achieve this goal, we must offer our students many opportunities to discuss their strategies and to recognize the relationships among the numbers in any addition or subtraction equation. As with any goal, it should serve as our focus as we steer students' talk about mathematics (Kazemi and Hintz 2014). Providing tasks that highlight such relationships gives students something to discuss.

In this game students draw cards randomly and write each number in an empty space on the *It's Greater* game board. The spaces represent digits in two-digit numbers, within expressions involving addition or subtraction. The goal is to have all sums and differences be as great as possible. In deciding where to write digits, therefore, players must think about the kind of numbers that will result in the greater sums or differences.

Math Focus

› Adding and subtracting two-digit numbers
› Understanding where to place digits in addition or subtraction examples to get greater sums or differences

Materials Needed

› 1 *It's Greater* Game Board per team (page A-71)
› 1 deck of *It's Greater* Cards, made from 2 copies of page A-72, per group
› Optional: 1 *It's Greater* Directions per group (page A-73)

> **Directions**
> Goal: Place digits in the spaces of the game board to create expressions with the greater sums and differences.
> - Shuffle the cards and place them facedown in a deck.
> - Turn over the top card.
> - Each team separately decides in which of the twenty spaces on its board to write the number. Be sure to notice the four *Not Using* squares in which you can write numbers. Once you write a number, it cannot be changed. Then discard this card and turn over the next card.
> - Continue playing until you have filled all twenty spaces with a number.
> - Teams add and subtract to complete each of their equations.
> - Compare your answers to each problem. The team with the greater answer gets 1 point. The team with the most points wins.

How It Looks in the Classroom

One teacher introduces this game using the numbers 0–9 written on letter-size paper. She asks four volunteers to come to the front of the room and each take a number randomly. Next, she asks them to stand in pairs to represent two numbers. The students have chosen the digits 1, 3, 6, and 9. The students with the digits 1 and 9 stand together, as do the students with 3 and 6. The teacher says, "You show the numbers 19 and 36. What is their sum?" She tells the observing students to also find the sum.

The students confirm that the sum is 55, and several students agree that the easiest way to find it is to think of $19 + 1 + 35$. Then the teacher asks how they might arrange these digits to form two-digit numbers with a greater sum. She tells them to write a number sentence to show the numbers and their sum. Again, all students are expected to respond. After a brief amount of time, she asks students to share their number sentences. The students suggest the following sentences:

$$91 + 36 = 127$$
$$19 + 63 = 82$$
$$91 + 63 = 154$$

The teacher asks, "Which of these sums is greater than the other two sums?" They all agree it's 154. The teacher does not want to engage students in a further conversation right now. She knows that after playing the game, more students will be able to contribute ideas about how to create numbers that are likely to have a greater sum.

The teacher distributes copies of the *It's Greater* game board to partners and explains how to play, making particular note of the *Not Using* squares. She calls out digits for them to write on their game boards, reminding them that once a digit is recorded, it cannot be erased and written elsewhere. When they complete a game, she has partners compare their boards with another set of partners to see whose sums and differences are greater.

CHAPTER 7
Addition and Subtraction

The next day they will play the entire game in small groups (with two sets of partners in each group) and then discuss game strategy as a large group.

Tips from the Classroom
- Most students need to play a few rounds before they begin to think strategically. Provide the opportunity for such play, without leading their thinking.
- Some students might not recognize, or might forget, that some examples involve addition and others, subtraction. You may want to suggest that they highlight the operation signs.
- After a round of play, encourage students with different placements of numbers to share their thinking.

What to Look For
- Do some students appear to write numbers randomly? Do others seem to take quite a while to decide where to write a digit?
- Do students recognize that 0 cannot be the first digit in a two-digit number?
- What evidence is there that students plan ahead when they discuss where to write a number?
- Do students think about how regrouping will impact the sum or difference they will get?

Variations
- Allow for a second round of play, when students get to change where they placed six of the numbers and refigure their answers and scores.
- Have players play *It's Less*, trying to get the sums and differences that are less.
- Make decks from three copies of page A-72 and have students play with game boards that have spaces for three-digit numbers and six *Not Using* spaces.

Exit Question Choices
- What reasons did you have for placing numbers in the *Not Using* spaces?
- How does thinking about place value help you place the digits?

Extension
Have students respond to the following journal-writing task.

You have four digits to place in the following subtraction problem. Each digit is different. Tell where to write the digits to get the greatest difference. Tell where to write the digits to get the least difference. Explain your thinking.

☐☐ − ☐☐ = _____

Meet the Rules

Why This Game or Puzzle?

Learning to calculate includes much more than procedural knowledge; it requires the understanding of relationships among numbers and mathematical thinking (Van den Heuvel-Panhuizen and Treffers 2009). When players must arrange digits to create expressions that meet a given condition or rule, students must decide which numbers to make and whether to add or subtract. When several different rules must be met, some students may note patterns, such as when one-, two-, or three-digit numbers are best.

In this game players roll five dice. The numbers on the dice identify the digits they may use to form one-, two-, or three-digit numbers. The players use the numbers, along with addition, subtraction, or both, to write an expression that meets a condition such as *The value is between 300 and 350.* The team that meets all game conditions first wins.

Math Focus

- Finding sums and differences
- Creating expressions with values within given ranges
- Looking for patterns and relationships among equations

Materials Needed

- 5 dice per group
- 1 *Meet the Rules* Game Board per team (page A-74)
- Optional: 1 *Meet the Rules* Directions per group (page A-75)

Directions

Goal: Roll dice and try to form expressions to meet the given rules before your opponent does.

- Take turns.
- On its first turn, Team 1 rolls five dice and talks about how to use each of the numbers shown once to meet the first rule. The team can make

> one-, two-, or three-digit numbers. The team can add or subtract or use both addition and subtraction. The answer must meet the rule.
> - If the team can use the numbers to fit the rule, it records the number sentence, has the other team check it, and gives the dice to the other team. On its next turn, Team 1 will try to fit the next rule.
> - If the team cannot use the numbers to fit the rule, its turn ends and it gives the dice to Team 2. On its next turn, Team 1 must roll and try to fit this same rule.
> - The first team to meet all five rules wins the game.

How It Looks in the Classroom

One second-grade teacher introduces this game to the whole class by demonstrating a turn. She explains that the game is about writing number sentences to fit rules. She writes the rule *It is between 75 and 90* on the whiteboard. She calls up two students and says, "Your task is to roll these dice and use each of the digits once to create one-, two-, or three-digit numbers. You can then add and/or subtract them. When you do, the answer should be between seventy-five and ninety. Please record what you roll on the whiteboard and talk aloud so that we will know what you are thinking."

One student rolls and reports the outcomes as the other student records the numbers on the board. They have rolled 2, 5, 2, 4, and 1. Quinn says, "We can't use the 4 and 5 as tens, because that would give us numbers that add to more than 90."

His partner, Rhonda, says, "Wait—let's try it. Maybe we can subtract the number that is left."

Quinn, says, "OK. Let's use the smaller numbers for the ones." He writes *41 + 52 = 93* and says, "We only have a 2 left. I wish we had a 4, 5, or 6 to subtract."

Rhonda says, "Let's try 2 and 5 in the tens." She writes *52 + 24 = 76*.

Quinn says, "Now we can just add the 1 and get seventy-seven."

The teacher is pleased with the focus on tens and ones. She explains the rest of the rules of the game, reminding students that they have to meet a rule on the game board, even if it takes several turns, before they can go on to the next rule.

Tips from the Classroom

- Some students may think about writing equations with only addition or subtraction. You might ask, *Is there another operation you could use? Could you use addition and subtraction in the same equation?*
- Players show interest in what their opponents have created and do tend to check for accuracy, though you may need to remind a few players of this expectation.
- Some students are more focused when time is restricted, while others are able to accomplish more without the pressure of a time limit. Offer the use of a timer and help students make the choice that is best for them.

› Occasionally, take time to review recording sheets with students so they know that the sheets give evidence of their thinking and are valued. While conferring with his teacher, the student who completed the recording shown in Figure 7.3 was quite excited to talk about how many times he rolled the same numbers. The teacher was impressed with his intuitive thinking about probability. The teacher also learned that the student did not notice that he needed to write a number sentence because he didn't think it mattered what the exact answer was, as long as it met the rule. The teacher asked him to complete the number sentences and check all of his work. When he did so, he also corrected his computation error in the last example.

What to Look For

› Do players' first attempts result in equations that meet or nearly meet the stated rule or ones that are significantly out of the given range of numbers?
› Do students talk about whether they want to create one-, two-, or three-digit numbers? What other conversations do you hear that you would like students to share with the larger group?
› What strategies do students use to find the values of the expressions they create?
› Do the players on a team work independently and then compare? Do they brainstorm ideas together?

Meet the Rules Game Board

Name(s): _____ Date: _____

1) The value is between 20 and 30. Number sentence:
   $26-3-3+3$

2) The value is between 75 and 90. Number sentence:
   $66+6+2+2$

3) The value is greater than 300 and less than 350. Number sentence:
   $351-1-1$

4) The value is between 80 and 100. Number sentence:
   $4+44-2$

5) The value is greater than 210 and less than 260. Number sentence:
   $236 + 6+6 = 256$

Figure 7.3 Sample student recording sheet

Variations

› Allow players to meet any one of the stated rules on a particular turn, instead of working on the rules in order.
› To vary difficulty levels, you could increase or decrease the number of dice students roll, the size of the numbers given in the rules, the number of rules students must meet, or the ranges of numbers given within the rules.

CHAPTER 7
Addition and Subtraction

Exit Question Choices
> What helped you to decide whether to create one-, two-, or three-digit numbers in your number sentences?
> You can use each of the digits 1, 2, 3, 4, and 5 once, along with addition, subtraction, or both, to write a number sentence with a value between 100 and 120. What is a number sentence you could write? (Possibilities include 142 − 35 = 107; 152 − 43 = 109; 153 − 42 = 111; 145 − 32 = 113; 123 − 4 − 5 = 114; 143 − 25 = 118).

Extension
Create a morning message that asks students to write a number sentence equal to 200 as they enter the classroom. Each sentence must include addition and subtraction. Once a sentence is recorded, it cannot be duplicated. Talk about any relationships students can identify within the equations. Are there any sentence pairs similar to 150 + 25 + 25 + 6 − 6 = 200 and 150 + 25 + 25 + 7 − 7 = 200 or 180 + 21 − 1 = 200 and 180 + 22 − 2 = 200?

Make Sense

Why This Game or Puzzle?
Children in the primary grades are capable of inductive and deductive reasoning (Stebbins 2003). Unfortunately, many young students are not presented with mathematical tasks that require logical thinking. Number puzzles often provide students with opportunities to think deductively and inductively as they develop their number sense.

Make Sense puzzles require solvers to place given digits within equations so that each equation makes mathematical sense. While students are likely to begin with a guess-and-check strategy, as their guessing continues, they might note patterns and generalizations such as if two one-digit numbers sum to a two-digit number, the tens digit in that sum must be one.

Math Focus
› Creating addition and subtraction equations
› Finding relationships and making generalizations
› Using inductive and deductive reasoning

Make Sense Puzzle A

Name(s): _____ Date: _____

Place each number in only one space so that the puzzle makes sense.

| 1, 2, 3, 4, 5, 6, 7, 8 |

☐ + ☐ = **3**

☐ + ☐ = **10**

☐ − ☐ = **4**

☐ − ☐ = **3**

> **Materials Needed**
> » 1 *Make Sense* Puzzle (A or B) per team (page A-76 or A-77)
> » Optional: 1 *Make Sense* Directions per team (page A-78)
>
> **Directions**
> Goal: Place the given numbers so that all of the equations in the puzzle make sense.
> » Work as a team.
> » Write each of the given numbers in only one space in the puzzle.
> » Check the math. Do all of the equations make sense? If not, try again.
> » Check your solution with another team.

How It Looks in the Classroom

One teacher introduces *Make Sense* puzzles by displaying the simpler one shown in Figure 7.4. She tells students to copy it and to think about it independently before sharing their thinking with a partner. She asks them to take some notes so they can remember their ideas.

> Place each number only once so that the puzzle makes sense.
>
> **1, 2, 3, 6, 8, 9**
>
> ☐ + ☐ = 10
>
> 4 − ☐ = ☐
>
> ☐ − 3 = ☐
>
> (Answer: 2 + 8 or 8 + 2 = 10; 4 − 1 = 3 or 4 − 3 = 1; 9 − 3 = 6)

Figure 7.4 Simpler *Make Sense* puzzle

Once an appropriate amount of time has passed, the teacher invites them to turn and talk to a neighbor. She walks around the classroom to listen as they talk and to identify students whom she wants to have share their thinking with the larger group. She notes some students' frustration when they place 1 and 9 in the first equation and then find the other numbers don't work in the spaces that remain. She reminds a few students that they can use the 2 only once. She smiles when she hears Suzanne exclaim, "The 3 and 1 have to go here. No other numbers work!" She observes a few students recognize that the numbers in the addition sentence could be reversed. She has several of these students share what they learned with the larger group.

Next the teacher tells them that they are going to work in partner teams to solve another *Make Sense* puzzle. Just before distributing copies of the puzzle, she says, "Be sure to think about some of the ideas we discussed. If you get stuck, try another possible solution, and look for numbers that must be placed in a particular space."

Tips from the Classroom

- In our field testing, most students began by making random choices and then realized that their choices were not going to allow them to complete each number sentence in a way that made sense. As they persevered, they began to recognize examples of where particular digits had to be placed; for example, in Puzzle A, only the numbers 1 and 2 have a sum of 3.
- Some students became frustrated with the number of times they had to erase. It helped them to write the numbers on small pieces of paper or plastic discs that they could move around as they considered alternatives.
- Encourage conversations as partner teams work together to find solutions, and require each team member to be able to explain the pair's thinking.
- Allow some students to work alone if they find it too challenging to solve the puzzle as part of a team, or provide all students some independent thinking time before they join their partners.

What to Look For

- What language do students use to describe their thinking? Do they refer to place values, addends, sums, and differences?
- What problem-solving strategies do students use? Do they guess and check, make lists, or use the process of elimination?
- Were some number sentences easier or more challenging than others?

Variations

- Provide only the digits necessary to solve two or three number sentences and include only those sentences in the puzzle.
- Have students play this as a game with teams taking turns. On each turn, a team places one of the digits and initials it. The other team can accept it or challenge it. A challenge means that the other team does not believe there are digits still available to complete the number sentence so it makes sense. If the challenging team is correct, the first team erases the digit and loses its turn. If the challenging team is incorrect, it loses its next turn. Play continues until neither team believes another digit can be placed or they have completely solved the puzzle. The team that placed more digits wins.

Exit Question Choices

› If you put a digit in each space shown below, so that the math is correct, what digit would you write in the tens place of the two-digit number? Why?

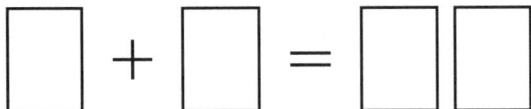

› What problem-solving strategies did you use to complete the *Make Sense* puzzle? Give an example of how these strategies helped you make one of the matches.

Extension

Have students complete a *What's the Problem?* puzzle, in which all of the missing digits can be determined by replacing letters with numbers to make the equation work. See Figure 7.5 for such a puzzle. Some students may also like to create such puzzles.

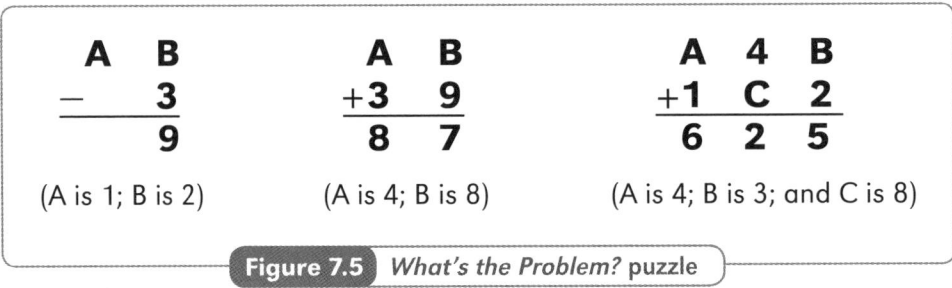

Figure 7.5 *What's the Problem?* puzzle

Online Games and Apps

Online games and apps may allow students to use both addition and subtraction while giving players the opportunity to make a variety of choices, including decisions related to number ranges, timed or untimed modes, and models to represent addition and subtraction strategies. Rather than simply let students practice procedural skills related to addition and subtraction, such games deepen students' understanding of these two operations and the relationships between them. A few examples include the following:

- Number Line Bounce, a free puzzle from the National Library of Virtual Manipulatives at http://nlvm.usu.edu/en/nav/frames_asid_107_g_1_t_1.html?from=category_g_1_t_1.html, engages students in addition and subtraction through a series of number line puzzles. For example, students may be given the number 3 to reach on the number line while being shown four bounce arrows of varying lengths, in this case, 5, 4, 3, and 1. Each time that the puzzler places a bounce arrow on the number line, he or she may choose to have the arrow move forward or backward on the number line. All bounce arrows must be used in order to reach the target number. Once the puzzler places the bounce arrows correctly and reaches the target number, the next step is to record a number sentence to represent the correct puzzle moves.

- Quento, a free app available at http://www.quento.com/, engages players in number puzzles in which they must make an expression to create a given sum or difference. A grid with five numbers and two addition and two subtraction signs is given, along with a target number. For example, puzzlers may be given the number 8 as the target and must choose from two (in the easier puzzles) or three of the numbers 5, 3, 7, 1, and 4 to create the target. The puzzles become increasingly difficult as students progress. Puzzlers who want to use four numbers to solve the puzzles must pay a fee to unlock those levels.

- Rock Hopper, a free game from Houghton Mifflin Math at http://www.eduplace.com/kids/mw/swfs/rockhopper_grade2.html, shows a frog looking at nine stepping-stones labeled with single-digit numbers. On some levels the numbers have addition or subtraction signs to accompany them. At the top of the screen, a large rock displays the target number. The players work out a route by jumping from rock to rock until the frog reaches the target. Players may choose different routes to reach the target, allowing for a good deal of mental math.

APPENDIX

Manners Expert Cards

Dear Manners Expert, Sometimes my math partner and I get too silly when we play math games. I'm not sure we should be partners. How could I choose another partner without hurting my friend's feelings?	Dear Manners Expert, My partner solved the math problems before I had a chance. He said he was just trying to help me, but I think I need to do the work, too, so I can learn. What should I do?
Dear Manners Expert, The other team got really mad today when they were losing the game. Sometimes we felt like we should try to lose so they would feel better. What can we say to help them not get so mad?	Dear Manners Expert, I am very shy and get nervous when we have to work in partners. I don't even know how to ask someone to be my partner. How can you help me?
Dear Manners Expert, Sometimes my math partner gives up when we get stuck. We tried to solve a puzzle today and it was tricky. She just stopped trying to solve it. How can I help her not give up?	Dear Manners Expert, My teammates don't always clean up. They throw things in the box and leave stuff on the floor. I am tired of being the one to put everything away the right way. What should I do?
Dear Manners Expert, Sometimes I really need to solve a puzzle alone. I get too distracted in the group. I think my teacher might let me, but I'm worried about what everyone would think. What should I do?	Dear Manners Expert, Yesterday my partner had some good ideas, but our turns took forever because he wouldn't make a choice. How can we get the other team to be more patient and my partner to be faster?
Dear Manners Expert, I'm afraid to get the answers wrong and so I just say I don't know. I know my partners get frustrated and think I don't know anything. What should I do?	Dear Manners Expert, I get so excited when I win that I clap and make a lot of noise. My math partner told me I was bragging and made the other team feel bad. How can I celebrate without bragging?
Dear Manners Expert, My math partner got a wrong answer today. When I told her, she got upset with me. I didn't mean to hurt her feelings. What should I do next time?	Dear Manners Expert, My partner and I had a lot of questions about the puzzle today, but everyone we could ask looked too busy. So we just filled in numbers. What do you think we should have done?

Well Played: Building Mathematical Thinking Through Number Games and Puzzles, Grades K–2
by Linda Dacey, Karen Gartland, and Jayne Bamford Lynch. Copyright © 2016. Stenhouse Publishers.

Count 20 Game Board

1	2	3	4	5
20				6
19				7
18				8
17				9
16				10
15	14	13	12	11

Count 20 Directions

Materials Needed
- 20 small counters, 10 in each of 2 colors, per group
- 1 die per group
- 1 cup per team
- 1 *Count 20* Game Board per group (page A-4)
- Optional: 1 *Count 20* Directions per group

Directions
Goal: Have the most counters in your cup at the end of the game.

- Decide which team goes first. The other team chooses the color of counters for each team.
- Each team begins with ten counters of the same color.
- On each turn:
  - Roll the die and choose a counter to move.
  - Count forward the number of spaces shown on the die. As one team member moves the counter, the other says the numbers on the spaces aloud. If there is another counter of either color on the number at which you finish, put that counter into your cup and leave your counter in that space.
  - If your move ends on 10, put your counter into your cup.
  - If your counter gets to 20, put it into your cup.
- The game ends when one of the teams does not have any counters to move.
- The team with the most counters in its cup wins.

Well Played: Building Mathematical Thinking Through Number Games and Puzzles, Grades K–2
by Linda Dacey, Karen Gartland, and Jayne Bamford Lynch. Copyright © 2016. Stenhouse Publishers.

Number Jigsaw Puzzle A

	5	6
	two	
10		4
	nine	one
seven	7	
		eight
3	three	5

Number Jigsaw Puzzle B

6	10	1 more than 10
	14	
8	1 less than 15	11
5	9	4
17	12	11
	1 more than 16 14	1 more than 13

Number Jigsaw Directions

Materials Needed
› 1 *Number Jigsaw* Puzzle (A or B) per pair (page A-6 or A-7)
› Optional: 1 *Number Jigsaw* Directions per pair

Directions
Goal: Arrange the puzzle pieces so that the numbers shown on all touching sides match.

› Work together.
› Place the nine puzzle pieces together to make a square.
› The numbers shown on all touching sides of the puzzle pieces must match.
› Check to make sure you have matched each side correctly.

Well Played: Building Mathematical Thinking Through Number Games and Puzzles, Grades K–2
by Linda Dacey, Karen Gartland, and Jayne Bamford Lynch. Copyright © 2016. Stenhouse Publishers.

Nim Recording Sheet

Name(s): _____   **Date:** _____

How many counters did Team 1 take? _____
How many counters are left? _____

How many counters did Team 2 take? _____
How many counters are left? _____

How many counters did Team 1 take? _____
How many counters are left? _____

How many counters did Team 2 take? _____
How many counters are left? _____

How many counters did Team 1 take? _____
How many counters are left? _____

How many counters did Team 2 take? _____
How many counters are left? _____

How many counters did Team 1 take? _____
How many counters are left? _____

How many counters did Team 2 take? _____
How many counters are left? _____

Well Played: Building Mathematical Thinking Through Number Games and Puzzles, Grades K–2
by Linda Dacey, Karen Gartland, and Jayne Bamford Lynch. Copyright © 2016. Stenhouse Publishers.

Nim Directions

Materials Needed
- 15 counters per group
- 2 *Nim* Recording Sheets per team (page A-9)
- Optional: 1 *Nim* Directions per group

Directions
Goal: Take the last counter.

- Place fifteen counters in a row.
- Decide who goes first. This is Team 1. The other team is Team 2.
- Take turns.
- On each turn, take one or two counters.
- After each turn, both teams write the number of counters taken and the number of counters left on their recording sheet.
- The team that takes the last counter wins.
- Play again. This time, Team 2 goes first.

Well Played: Building Mathematical Thinking Through Number Games and Puzzles, Grades K–2 by Linda Dacey, Karen Gartland, and Jayne Bamford Lynch. Copyright © 2016. Stenhouse Publishers.

Mystery Number Recording Sheet

Name(s): _____   Date: _____

The mystery number has _____ digits.

Questions Asked	Yes or No?

Well Played: Building Mathematical Thinking Through Number Games and Puzzles, Grades K–2
by Linda Dacey, Karen Gartland, and Jayne Bamford Lynch. Copyright © 2016. Stenhouse Publishers.

Mystery Number Directions

Materials Needed
> Optional: 1 *Mystery Number* Recording Sheet per group (page A-11)
> Optional: 1 *Mystery Number* Directions per group

Directions
Goal: Name the mystery number.

> Choose two game leaders.
> The game leaders write a number and keep it private. They tell how many digits are in the number.
> The other players take turns asking yes-or-no questions. The leaders answer.
> Players record questions and answers on the recording sheet.
> Ask questions until a player names the mystery number by asking, *Is the mystery number* _____?
> The leaders show their written number when a player names it correctly. If players make an incorrect guess, they should continue asking questions until they make a correct one.

Order Up Cards

7	8	17	18
24	26	29	30
31	42	45	56

Well Played: Building Mathematical Thinking Through Number Games and Puzzles, Grades K–2
by Linda Dacey, Karen Gartland, and Jayne Bamford Lynch. Copyright © 2016. Stenhouse Publishers.

Order Up Cards (continued)

59	60	63	64
78	79	81	84
90	97	98	99

Order Up Cards (optional)

100	102	115	124
134	138	151	158
167	170	182	199

Well Played: Building Mathematical Thinking Through Number Games and Puzzles, Grades K–2
by Linda Dacey, Karen Gartland, and Jayne Bamford Lynch. Copyright © 2016. Stenhouse Publishers.

Order Up Directions

Materials Needed
› 1 deck of *Order Up* Cards per group (pages A-13–A-15)
› Optional: 1 *Order Up* Directions per group

Directions
Goal: Put five numbers in order from least to greatest, left to right.

› Shuffle the cards and place five cards faceup, from left to right, in front of each team. Teams may not change the order of the cards in their "hands."
› Put the other cards facedown in a deck.
› Choose a player from each team to do rock-paper-scissors. The winning player's team goes *second.*
› The first team chooses the top card from the deck. The team may trade this card for one of its five cards. If the team makes a trade, it places the card from its hand faceup next to the deck in a discard pile. If the team does not make a trade, it places the card drawn from the deck faceup in the discard pile.
› The players take turns, picking a card from either the deck or the discard pile. Then they place either a card from their hand or the card that was drawn faceup on top of the cards next to the deck.
› The first team to get its five numbers in order from least to greatest, left to right, is the winner.

Win 1,000 Place-Value Chart

Ones	
Tens	
Hundreds	
Thousands	

Well Played: Building Mathematical Thinking Through Number Games and Puzzles, Grades K–2
by Linda Dacey, Karen Gartland, and Jayne Bamford Lynch. Copyright © 2016. Stenhouse Publishers.

Win 1,000 Directions

Materials Needed
- 2 dice, each a different color, per group
- 1 *Win 1,000* Place-Value Chart per team (page A-17)
- A collection of base ten materials per group
- Optional: 1 *Win 1,000* Directions per group

Directions
Goal: Show 1,000 (or more) on your place-value chart.

- One color die represents tens, and the other die represents ones. It is different for each team. For example, if Team 1 chooses red for ones, red represents tens for Team 2.
- Teams take turns rolling the dice.
- Team 1 looks at the dice and places that number of tens and ones on its place-value chart. If it can make a trade for a greater place value, it must do so. For example, if it has 10 ones, it must trade them for 1 ten. Team 1 reads the number represented on its chart.
- After the players on Team 2 agree with the number Team 1 read, they use the same roll of the dice to place their number of tens and ones on their place-value chart, trade if they can, and read their number to Team 1.
- The first team to show 1,000 or more on its place-value chart wins.

Well Played: Building Mathematical Thinking Through Number Games and Puzzles, Grades K–2
by Linda Dacey, Karen Gartland, and Jayne Bamford Lynch. Copyright © 2016. Stenhouse Publishers.

Number Sort Label Cards

Less than 10	Greater than ten
Every digit in the ones place is the same	Every digit in the ones place is different
Has a 2 in the tens place	Has a 1 in the tens place
Greater than 20	Less than 20
The digit in the ones place is greater than the digit in the tens place	The digit in the tens place is greater than the digit in the ones place
Odd number	Even number

Well Played: Building Mathematical Thinking Through Number Games and Puzzles, Grades K–2 by Linda Dacey, Karen Gartland, and Jayne Bamford Lynch. Copyright © 2016. Stenhouse Publishers.

Number Sort Directions

Materials Needed
- 2 large strings tied in loops or 1 large Venn diagram drawn on chart paper per group
- 35 chips, labeled with the numbers 1–35, per group
- 1 deck of *Number Sort* Label Cards per group (page A-19)
- Optional: 1 *Number Sort* Directions per group

Directions
Goal: Guess the two sorting rules.

- Decide which team is Team 1. The other team is Team 2.
- Team 1 looks at the cards and chooses two labels, one for each ring of the diagram. Team 1 places each label facedown beside its correct ring and does not tell Team 2 what the labels are.
- Team 2 needs to guess the labels. It chooses a numbered chip and gives it to Team 1. Team 1 correctly places the numbered chip in the diagram.
- Play continues until Team 2 correctly guesses the labels.
- Team 2 must guess both labels at once. Team 2 should also explain its thinking when it makes a guess.
- If Team 2 guesses one label right and one label wrong, Team 1 just says, "No." Team 1 doesn't give any hints about which guess is right or wrong.
- When Team 2 guesses the right label for each ring, Team 1 shows the labels.
- Have teams change roles and play another game.

Well Played: Building Mathematical Thinking Through Number Games and Puzzles, Grades K–2 by Linda Dacey, Karen Gartland, and Jayne Bamford Lynch. Copyright © 2016. Stenhouse Publishers.

Go Number Fish Cards

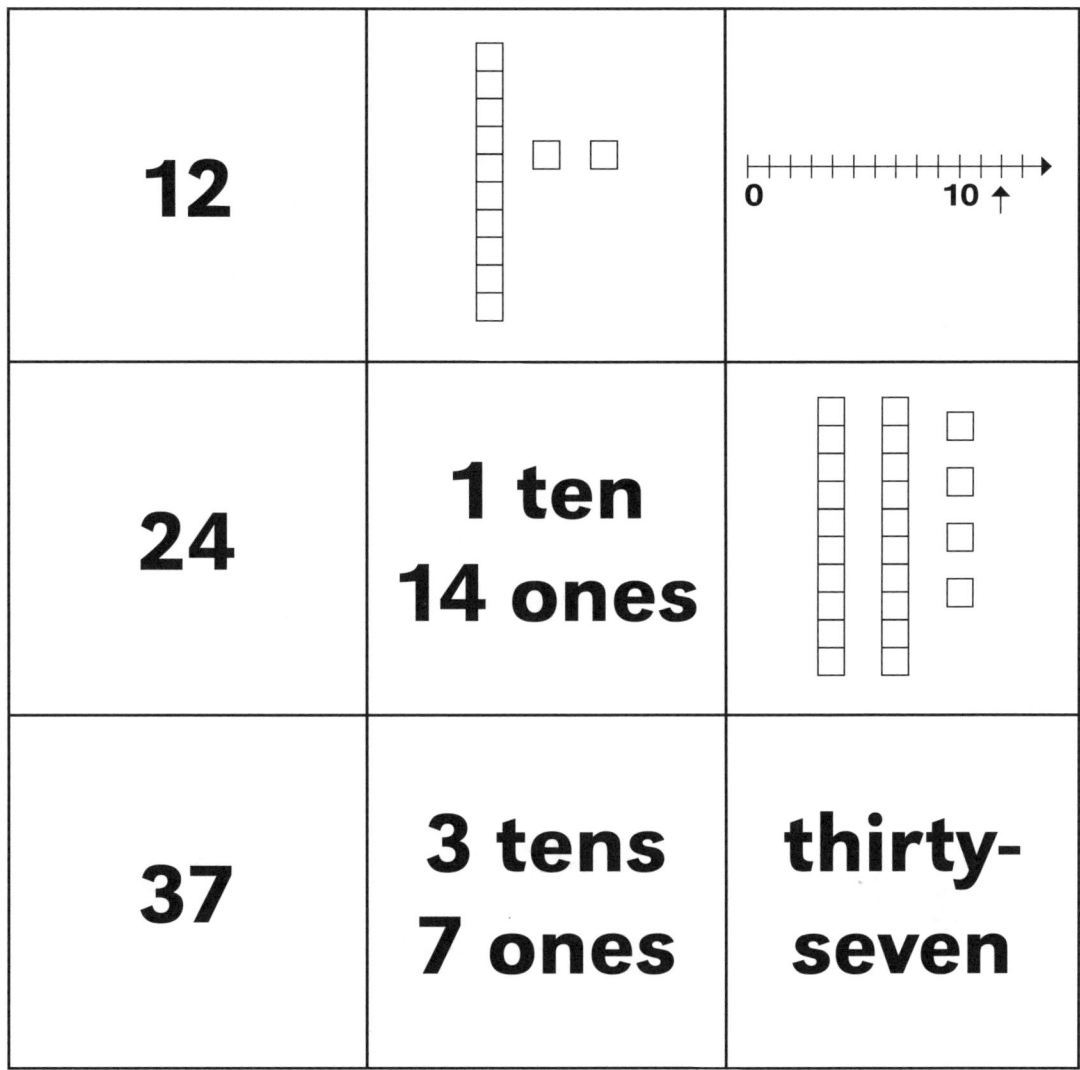

Go Number Fish Cards (continued)

40	**3 tens 10 ones**	
57	**5 tens 7 ones**	
61	**5 tens 11 ones**	**sixty-one**

Go Number Fish Cards (continued)

75	7 tens 5 ones	6 tens 15 ones
83	8 tens 3 ones	*number line from 80 to 90 with arrow pointing to 83*
96	8 tens 16 ones	ninety-six

Go Number Fish Directions

Materials Needed
- 1 file folder per team
- 1 deck of *Go Number Fish* Cards per group (pages A-21–A-23)
- Optional: 1 copy of *Go Number Fish* Directions per group

Directions
Goal: Make the greater number of packs of cards.

- Decide who goes first.
- Mix up the cards. Deal five cards to each team.
- Teams place their cards behind a standing file folder so the other team can't see them.
- Place the other cards facedown in a deck.
- Take turns.
- On each turn, choose a card behind your folder and ask if the other team has a card with the same value. You might ask, "Do you have a card with a value of twelve?"
- The other team must give you any cards it has that match what you asked for, and then you get another turn.
- If the team does not have such a card, it tells you to "go number fish." You fish by taking the top card from the deck.
- If the card has the value you asked for, show it, put it in your hand, and take another turn. If not, your turn is over.
- When you get three cards with the same value, you have made a pack. Place each pack faceup in front of your folder.
- The game ends when a team has no cards left.
- The team with the greater number of packs wins.

The Number Is/What Number Is? Cards A

The number is 80. What number is 1 ten more than 19?	The number is 29. What number is 7 tens and 3 ones?	The number is 73. What number is 1 ten less than 6 tens?
The number is 50. What number is 1 ten more than 32?	The number is 42. What number is 6 tens and 8 ones?	The number is 68. What number is 1 ten less than 40?
The number is 30. What number is 8 tens and 9 ones?	The number is 89. What number is 1 ten more than 49?	The number is 59. What number is 9 tens and 7 ones?
The number is 97. What number is 1 ten less than 2 tens?	The number is 10. What number is 2 tens and 5 ones?	The number is 25. What number is 10 more than 7 tens?

Well Played: Building Mathematical Thinking Through Number Games and Puzzles, Grades K–2
by Linda Dacey, Karen Gartland, and Jayne Bamford Lynch. Copyright © 2016. Stenhouse Publishers.

The Number Is/What Number Is? Cards B

The number is 200. What number is 1 ten more than 9 tens?	The number is 100. What number is 2 hundreds, 6 tens, and 5 ones?	The number is 265. What number is 1 less than 30 tens?
The number is 299. What number is 5 ones and 6 hundreds?	The number is 605. What number is 300 + 60 + 3?	The number is 363. What number is 1 ten less than 400?
The number is 390. What number is 10 more than 520?	The number is 530. What number is 200 more than 99?	The number is 299. What number is 1 ten less than 7 hundreds?
The number is 690. What number is 36 tens?	The number is 360. What number is 4 tens and 8 hundreds?	The number is 840. What number is 20 tens?

Well Played: Building Mathematical Thinking Through Number Games and Puzzles, Grades K–2 by Linda Dacey, Karen Gartland, and Jayne Bamford Lynch. Copyright © 2016. Stenhouse Publishers.

The Number Is/What Number Is? Directions

Materials Needed
- 1 deck of *The Number Is/What Number Is?* Cards (A or B) per group (page A-25 or A-26)
- Optional: 1 *The Number Is/What Number Is?* Directions per group

Directions
Goal: Place cards so that the number identified on each card answers the question on the card before it.

- Mix up the cards and place them faceup on a table or the floor.
- Choose a card and read its question.
- Find a card with a matching answer.
- Place this card next to the first card.
- Read the question on this second card. Find a card with a matching answer and place it next to the second card.
- Continue to read questions and find answers. Put the cards in a circle so that each question is followed with a correct answer.
- Each card must fit in the circle.

Well Played: Building Mathematical Thinking Through Number Games and Puzzles, Grades K–2 by Linda Dacey, Karen Gartland, and Jayne Bamford Lynch. Copyright © 2016. Stenhouse Publishers.

Number Touch Game Board

451	270	382	95	247	81
302	194	45	148	536	253
36	29	468	379	50	79
527	310	73	92	487	14
104	265	90	536	61	128
49	386	58	103	279	75
413	97	152	258	304	580
260	74	521	85	410	93

Well Played: Building Mathematical Thinking Through Number Games and Puzzles, Grades K–2 by Linda Dacey, Karen Gartland, and Jayne Bamford Lynch. Copyright © 2016. Stenhouse Publishers.

Number Touch Recording Sheet

Name(s): _____ Date: _____

Turn	Team 1 Total Score	Team 2 Total Score
1		
2		
3		
4		
5		
6		
7		
8		

Well Played: Building Mathematical Thinking Through Number Games and Puzzles, Grades K–2
by Linda Dacey, Karen Gartland, and Jayne Bamford Lynch. Copyright © 2016. Stenhouse Publishers.

Number Touch Digit Cards

0	1	2	3
4	5	6	7
8	9		

Well Played: Building Mathematical Thinking Through Number Games and Puzzles, Grades K–2
by Linda Dacey, Karen Gartland, and Jayne Bamford Lynch. Copyright © 2016. Stenhouse Publishers.

Number Touch Directions

Materials Needed
- 1 *Number Touch* Game Board per group (page A-28)
- About 10 each of tongue depressor sticks labeled *hundreds, tens,* and *ones,* mixed up and placed in an envelope, per group
- 1 deck of *Number Touch* Digit Cards per group (page A-30)
- 1 *Number Touch* Recording Sheet per group (page A-29)
- Optional: 1 *Number Touch* Directions per group

Directions
Goal: Earn the greater number of points by finding numbers on the game board and earning points for those numbers and any number they touch.

- Use rock-paper-scissors to decide which team goes first. The winner goes *second.*
- Place the cards facedown. Move them around so they are shuffled.
- Teams take turns. On each turn, choose a card and a stick. Look for a number on the board with the digit on the card in the place on the stick. Mark it with an X. This number may not be marked again. For example, if you draw the card with the 2 and a stick labeled *tens,* you should look for a number with a 2 in the tens place.
- You get 1 point for finding a correct unmarked number on the board and 1 point for each other *marked* number next to it (in a row, column, or diagonal). If you could mark 148 on the mini-board shown below, you would receive 3 points: 1 point for 148, 1 for touching 536, and 1 for touching 247.
- If you cannot mark a card, your turn is over.
- Each team gets eight turns.

382	95	2̶4̶7̶
45	148	5̶3̶6̶
468	379	50

- Write your score on the recording sheet. After the first turn, be sure to add your old score to your new one before you write. The team with the greater score at the end wins.

Well Played: Building Mathematical Thinking Through Number Games and Puzzles, Grades K–2
by Linda Dacey, Karen Gartland, and Jayne Bamford Lynch. Copyright © 2016. Stenhouse Publishers.

Make a Pair Five-Frame Cards

Make two copies of these cards to form one deck.

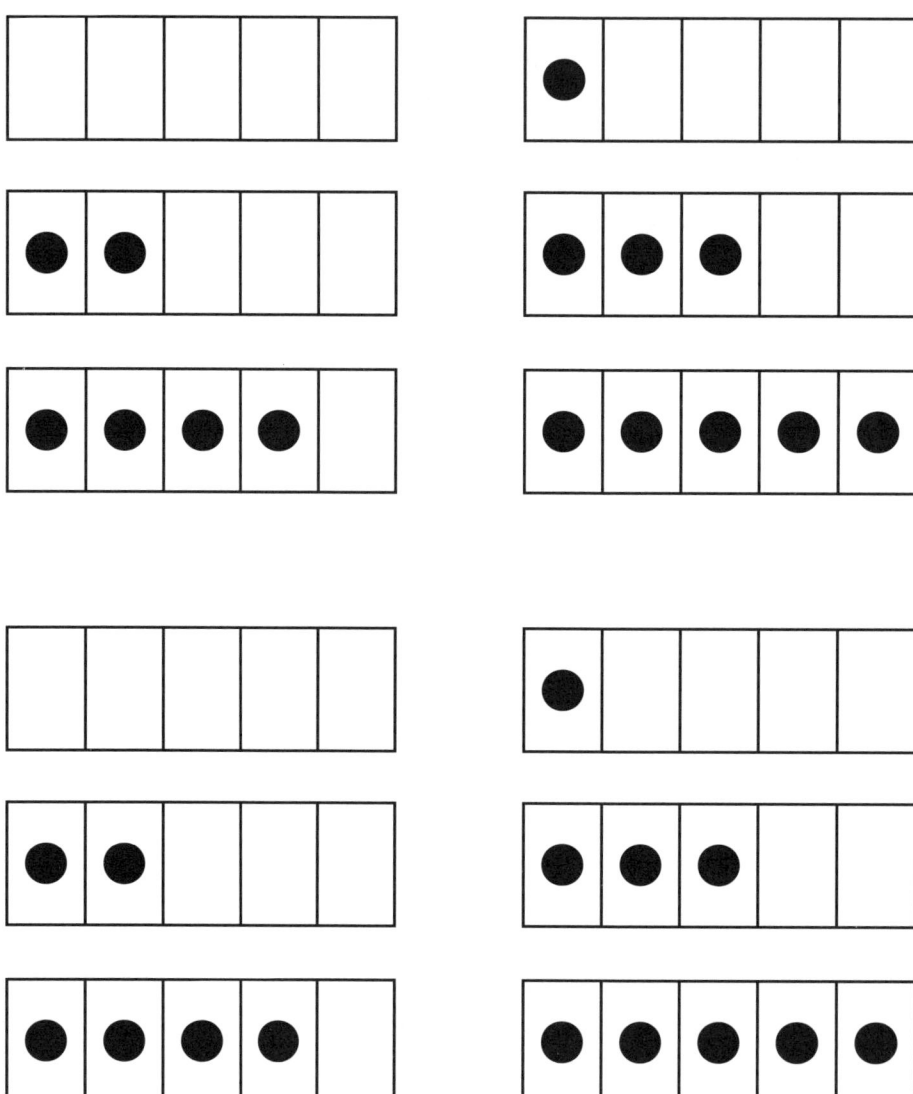

Make a Pair Ten-Frame Cards

Make four copies of these cards to form one deck.

Make a Pair Directions

Materials Needed
- 1 deck of *Make a Pair* Five-Frame Cards, made from 2 copies of page A-32 per group
- Optional: 5 counters per team
- Optional: 1 *Make a Pair* Directions per group

Directions
Goal: Be the first team to find five pairs of five-frame cards with a total of five dots.

- Mix up the cards. Put them facedown in a pile.
- Each team takes three cards from the pile.
- Turn over the top card of the pile and place it faceup beside the pile.
- Take turns.
- On each turn you can do one of three things:
  1. Put two of your cards down, if you think they have a total of five dots. If the other team agrees, you have a pair.
  2. Pick up the card that is faceup and put it down along with one of your cards, if you think the total number of dots is five. If the other team agrees, you have a pair.
  3. Take the top card of the pile and see if you can make a pair. If the other team agrees, you have a pair. If you do not have a pair, keep the card for use next time.
- After each turn, if you do not have at least three cards, pick up cards from the pile until you have three cards in your hand. Also, turn over a new card from the pile, if one is not shown faceup already.
- The game ends when a team makes five pairs. The first team with five pairs wins.

Equal Values Cards

=	=
2 + 3	3 + 2
4 + 1	5 + 0
0 + 4	2 + 2
2 + 1	1 + 2
1 + 1	2 + 0
1 + 0	0 + 1

Well Played: Building Mathematical Thinking Through Number Games and Puzzles, Grades K–2
by Linda Dacey, Karen Gartland, and Jayne Bamford Lynch. Copyright © 2016. Stenhouse Publishers.

Equal Values Cards (continued)

2 + 4	3 + 3
4 + 3	2 + 5
4 + 5	5 + 4
2 + 6	6 + 2
3 + 7	2 + 8
6 + 4	5 + 5
1 + 9	0 + 10

Equal Values Cards (continued)

6 + 5	5 + 6
9 + 3	6 + 6
6 + 7	7 + 6
7 + 7	5 + 9
9 + 6	8 + 7
7 + 9	8 + 8

Well Played: Building Mathematical Thinking Through Number Games and Puzzles, Grades K–2
by Linda Dacey, Karen Gartland, and Jayne Bamford Lynch. Copyright © 2016. Stenhouse Publishers.

Equal Values Recording Sheet

Name(s): _____   Date: _____

_____ = _____

_____ = _____

_____ = _____

_____ = _____

_____ = _____

_____ = _____

_____ = _____

_____ = _____

_____ = _____

Well Played: Building Mathematical Thinking Through Number Games and Puzzles, Grades K–2 by Linda Dacey, Karen Gartland, and Jayne Bamford Lynch. Copyright © 2016. Stenhouse Publishers.

Equal Values Directions

Materials Needed
- 1 deck of *Equal Values* Cards per group (pages A-35–A-37)
- 1 *Equal Values* Recording Sheet per team (page A-38)
- Optional: 1 *Equal Values* Directions per group

Directions
Goal: Get the most pairs of cards that have equal values.

- Give each team a card with the equal sign.
- Shuffle the remaining cards. Deal each team four cards faceup for all to see. Put the other cards facedown in a pile.
- Decide which team goes first.
- On each turn, you can do one of three things:
  1. Find two of your cards that have an equal value. Set this pair beside you. Replace them with two cards from the top of the pile.
  2. Trade one of your cards with one of the other team's cards when that lets you make a pair. Set this pair beside you. Replace your card with a card from the top of the pile.
  3. Draw a card from the top of the deck and add it to your cards.
- When a team makes a pair, both teams must agree that the sums are equal and then the team that made the pair must record the expressions on its recording sheet.
- If no cards are left in the pile, you can still have a turn, but you can't take a card from the pile.
- The game ends when no team can make another pair.
- The team with more pairs wins.

Well Played: Building Mathematical Thinking Through Number Games and Puzzles, Grades K–2
by Linda Dacey, Karen Gartland, and Jayne Bamford Lynch. Copyright © 2016. Stenhouse Publishers.

Triangle Totals Puzzles Sheet A

Name(s): _____   Date: _____

Place the numbers 1, 2, 3, 4, 5, and 6 in the circles.
Use each number only once.
Add the numbers on each side.
The total should be 9 on each side.

Place the numbers 1, 2, 3, 4, 5, and 6 in the circles.
Use each number only once.
Add the numbers on each side.
The total should be 12 on each side.

Well Played: Building Mathematical Thinking Through Number Games and Puzzles, Grades K–2 by Linda Dacey, Karen Gartland, and Jayne Bamford Lynch. Copyright © 2016. Stenhouse Publishers.

Triangle Totals Puzzles Sheet B

Name(s): _____ Date: _____

Put the numbers 4, 5, 6, 7, 8, and 9 in the circles.
Use each number only once.
Add the numbers on each side.
The total should be 18 on each side.

Put the numbers 4, 5, 6, 7, 8, and 9 in the circles.
Use each number only once.
Add the numbers on each side.
The total should be 20 on each side.

Want more puzzles? Use these same numbers to get a total of 19 on each side. Then try for 21 on each side.

Well Played: Building Mathematical Thinking Through Number Games and Puzzles, Grades K–2
by Linda Dacey, Karen Gartland, and Jayne Bamford Lynch. Copyright © 2016. Stenhouse Publishers.

Triangle Totals Directions

Materials Needed
- 6 chips numbered *1–6* or *4–9* per team
- 1 *Triangle Totals* Puzzles Sheet (A or B) per team (page A-40 or A-41)
- Optional: 1 *Triangle Totals* Directions per team

Directions
Goal: Place the numbers in the triangle so that when you add the three numbers on each side of the triangle, you get the given total.

- Place each of the given numbers on the triangle.
- Add the numbers on each side.
- Check to see if each side matches the given total.
- If not, try again.
- Check your solution with another team.

Well Played: Building Mathematical Thinking Through Number Games and Puzzles, Grades K–2 by Linda Dacey, Karen Gartland, and Jayne Bamford Lynch. Copyright © 2016. Stenhouse Publishers.

Yahoo! 100 Cards

Make three copies of these cards to form one deck.

0	1	2	3
4	5	6	7
8	9	10	20
30	40	50	60
70	80	90	

Well Played: Building Mathematical Thinking Through Number Games and Puzzles, Grades K–2
by Linda Dacey, Karen Gartland, and Jayne Bamford Lynch. Copyright © 2016. Stenhouse Publishers.

Yahoo! 100 Recording Sheet

Name(s): _____ Date: _____

	Equation	Points for Turn	Total Points
Turn 1			
Turn 2			
Turn 3			
Turn 4			
Turn 5			
Turn 6			

Well Played: Building Mathematical Thinking Through Number Games and Puzzles, Grades K–2
by Linda Dacey, Karen Gartland, and Jayne Bamford Lynch. Copyright © 2016. Stenhouse Publishers.

Yahoo! 100 Directions

Materials Needed
- 1 deck of *Yahoo! 100* Cards, made from 3 copies of page A-43, per group
- 1 *Yahoo! 100* Recording Sheet per team (page A-44)
- Optional: 1 *Yahoo! 100* Directions per group

Directions
Goal: Earn the greater amount of points by collecting numbers to make sets with totals less than or equal to one hundred.

- Put the cards facedown between the teams. Spread them out and mix them up to shuffle them. Leave them spread out for the game.
- Take turns.
- Begin each turn by turning over two cards. Add the numbers. If you choose to turn over another card, you *must* also add that number to your total.
- Decide when to stop turning over a card and adding the number to your total.
- If your total is less than one hundred when you stop, you get 1 point for each card you used. If it is equal to one hundred, you get 1 point for each card and 10 bonus points. If it is greater than one hundred, you get 0 points.
- Record a number sentence for this turn, the points it is worth, and your total points for the game.
- Put the cards back facedown and mix them all up again before the other team takes its turn.
- The team with the greater total score after six rounds wins.

Well Played: Building Mathematical Thinking Through Number Games and Puzzles, Grades K–2
by Linda Dacey, Karen Gartland, and Jayne Bamford Lynch. Copyright © 2016. Stenhouse Publishers.

On Target Recording Sheet

Name(s): _____ Date: _____

The target sum for Round 1 is _____.

☐☐☐ + ☐☐☐ = _____

The target sum for Round 2 is _____.

☐☐☐ + ☐☐☐ = _____

The target sum for Round 3 is _____.

☐☐☐ + ☐☐☐ = _____

The target sum for Round 4 is _____.

☐☐☐ + ☐☐☐ = _____

The target sum for Round 5 is _____.

☐☐☐ + ☐☐☐ = _____

Well Played: Building Mathematical Thinking Through Number Games and Puzzles, Grades K–2
by Linda Dacey, Karen Gartland, and Jayne Bamford Lynch. Copyright © 2016. Stenhouse Publishers.

On Target Directions

Materials Needed
- 3 dice per group
- 1 deck of playing cards, with face cards removed, per group
- 1 *On Target* Recording Sheet per team (page A-46)
- Optional: 1 *On Target* Directions per group

Directions
Goal: Place digits in a number sentence to get as close as possible to a given sum.

- Mix up the cards.
- Give each team six cards.
- Roll the three dice, one at a time. The first one rolled shows hundreds, the second shows tens, and the third shows ones. This is the target sum. Write it on your recording sheets.
- Each team places its cards to form two three-digit numbers, writes the numbers on its recording sheet, and records the sum.
- Teams compare the sums. Circle the one that is closer to the target sum.
- Play five rounds. The team with more sums closer to the target numbers wins.

Well Played: Building Mathematical Thinking Through Number Games and Puzzles, Grades K–2
by Linda Dacey, Karen Gartland, and Jayne Bamford Lynch. Copyright © 2016. Stenhouse Publishers.

How Many Are in the Cup? Recording Sheet

Name(s): _____   Date: _____

1) _____ − _____ = _____

2) _____ − _____ = _____

3) _____ − _____ = _____

4) _____ − _____ = _____

5) _____ − _____ = _____

6) _____ − _____ = _____

Well Played: Building Mathematical Thinking Through Number Games and Puzzles, Grades K–2
by Linda Dacey, Karen Gartland, and Jayne Bamford Lynch. Copyright © 2016. Stenhouse Publishers.

How Many Are in the Cup? Directions

Materials Needed
- 5 or 10 counters per group
- 1 opaque cup per group
- 1 *How Many Are in the Cup?* Recording Sheet per group (page A-48)
- Optional: 5 or 10 counters per team for modeling
- Optional: 1 *How Many Are in the Cup?* Directions per group

Directions
Goal: Find the number of missing counters.
- Put the counters in the cup (either five or ten counters).
- Decide which team is Team 1. The other team is Team 2.
- Team 1 takes some of the counters out of the cup and puts them between the two teams. The teams agree how many counters they can all see.
- Team 2 decides how many counters are still in the cup.
- Team 1 counts the counters in the cup.
- Team 2 writes the subtraction number sentence on the recording sheet.
- Take turns being Team 1 and Team 2.
- The game ends when, together, the teams have written six number sentences.

Well Played: Building Mathematical Thinking Through Number Games and Puzzles, Grades K–2
by Linda Dacey, Karen Gartland, and Jayne Bamford Lynch. Copyright © 2016. Stenhouse Publishers.

Move Along Game Board

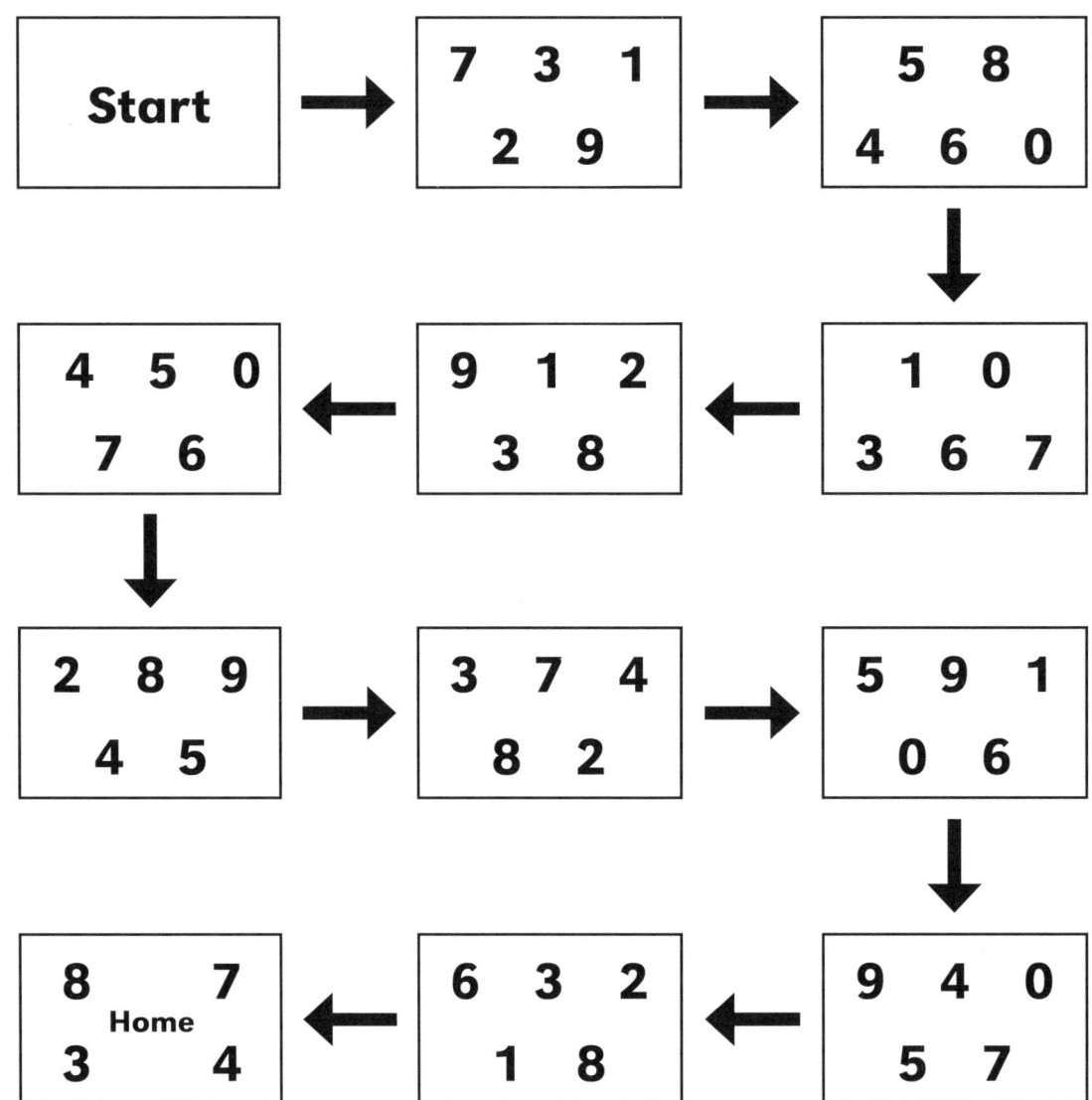

Move Along Cards

1	2	3	4
5	6	7	8
9	10		

Well Played: Building Mathematical Thinking Through Number Games and Puzzles, Grades K–2
by Linda Dacey, Karen Gartland, and Jayne Bamford Lynch. Copyright © 2016. Stenhouse Publishers.

Move Along Directions

Materials Needed
- 2 game pieces per group
- 1 *Move Along* Game Board per group (page A-50)
- 1 deck of *Move Along* Cards per group (page A-51)
- Optional: 1 *Move Along* Directions per group

Directions
Goal: Be the first team to land on the Home space.

- Each team puts its game piece on Start.
- Mix up the game cards and spread them out facedown between the teams.
- Take turns.
- On each turn:
  - Pick up two cards. Choose a number on one of the cards to subtract from ten.
  - If the answer is in the next space on the game board, move your team's piece to that space.
  - If the answer is not there, subtract the other number from ten.
  - If that answer is there, move your team's game piece to that space.
  - If neither answer is there, your team's game piece stays where it is, and your turn is over.
- After each turn, put your cards facedown and mix up all the cards.
- Whichever team makes it to the Home space first wins the game.

Take the Numbers Board Number Cards

11	12	13
14	15	16
17	18	19

Well Played: Building Mathematical Thinking Through Number Games and Puzzles, Grades K–2
by Linda Dacey, Karen Gartland, and Jayne Bamford Lynch. Copyright © 2016. Stenhouse Publishers.

Take the Numbers Recording Sheet

Name(s): _____   Date: _____

Each time you put a card below a game board number, subtract. Cross out the number shown and write the number that is left.

11	12	13	14	15	16	17	18	19

Take the Numbers Directions

Materials Needed
› 1 set of *Take the Numbers* Board Number Cards per group (page A-53)
› 1 deck of playing cards, without the face cards (aces stand for 1), per group
› 1 *Take the Numbers* Recording Sheet per group (page A-54)
› Optional: 1 *Take the Numbers* Directions per group

Directions
Goal: Collect the greater number of playing cards.

› Put the nine board numbers faceup in a row so everyone can see them.
› Mix up the playing cards. Put them facedown in a deck.
› Deal four playing cards to each team faceup for all to see.
› Take turns. On each turn you place one of your playing cards below one of the board numbers. You can put your playing card below a board number that already has a playing card there or below one that does not. Then look at the recording sheet for that number.
› If you put the first card below a number, subtract the value of your playing card from the board number you chose. Cross out the board number on the recording sheet and write the difference you found in that column. Then take a new playing card from the deck to add to your hand.
› When there is already a card below the board number you chose, check the recording sheet. If your card is less than the last number in the column on the recording sheet for the board number you chose, subtract your card from that number. Cross out the last number in that column and record the new difference.
› If your card is equal to the last number in the column, subtract and record the zero difference. Turn over that game board number, as it is no longer in play, and take all of its playing cards. Put them facedown near your team. Take a card from the deck to put in your hand. Your turn is over.
› If you can't place a number that gives a difference greater than or equal to zero, you lose your turn.
› The game ends when neither team can place a card. The team that has collected the greater number of playing cards wins.

Well Played: Building Mathematical Thinking Through Number Games and Puzzles, Grades K–2
by Linda Dacey, Karen Gartland, and Jayne Bamford Lynch. Copyright © 2016. Stenhouse Publishers.

Name That Number Clues A

The number is greater than 9 − 5.	The number is less than 10 − 1.
The number is not 10 − 4.	The number is not 8 − 0.
The number is not 9 − 2.	What is the number?

Name That Number Clues B

The number is greater than 100 − 100.	The number is less than 100 − 90.
The number is even.	The number is not 12 − 4.
The number is not 13 − 7.	The number is not 10 − 8. What is the number?

Well Played: Building Mathematical Thinking Through Number Games and Puzzles, Grades K–2 by Linda Dacey, Karen Gartland, and Jayne Bamford Lynch. Copyright © 2016. Stenhouse Publishers.

Name That Number Clues C

The number is greater than 570 − 430.	The number is less than 852 − 692.
All digits in the number are odd.	No digits are the same as in the answer to 564 − 85.
The number is not 782 − 627.	The number is not 472 − 321. What is the number?

Name That Number Directions

Materials Needed
- 1 set of *Name That Number* Clues (A, B, or C) per group (pages A-56, A-57, or A-58)
- Optional: 1 *Name That Number* Directions per group

Directions
Goal: Use the clues to find the mystery number.
- Work as a team of three puzzle solvers.
- Place the clues facedown. Each solver randomly takes two of the clues.
- Decide how to share the clues.
- Work together, read the clues as many times as necessary, and talk about what you know. Try to find the number that fits all the clues.
- When you think you have the solution, read the clues again to check.

Well Played: Building Mathematical Thinking Through Number Games and Puzzles, Grades K–2
by Linda Dacey, Karen Gartland, and Jayne Bamford Lynch. Copyright © 2016. Stenhouse Publishers.

Subtraction Tic-Tac-Toe Game Board A

Sign A

10	8
	9

Sign B

7	1
	4

Answer Board

3	4	8
7	2	6
5	9	1

Subtraction Tic-Tac-Toe Game Board B

Sign A

58	97
46	75

Sign B

30	10
40	5

Answer Board

28	45	87	6
41	67	35	48
65	18	16	92
57	36	53	70

Well Played: Building Mathematical Thinking Through Number Games and Puzzles, Grades K–2 by Linda Dacey, Karen Gartland, and Jayne Bamford Lynch. Copyright © 2016. Stenhouse Publishers.

Subtraction Tic-Tac-Toe Game Board C

Sign A

899	192
389	625

Sign B

81	16
188	141

Answer Board

883	308	484	4
248	711	176	544
437	51	818	373
111	609	201	758

Subtraction Tic-Tac-Toe Directions

Materials Needed
- 1 *Subtraction Tic-Tac-Toe* Game Board (A, B, or C) per group (page A-60, A-61, or A-62)
- Optional: 1 calculator per team
- Optional: 1 *Subtraction Tic-Tac-Toe* Directions per group

Directions
Goal: Choose pairs of numbers to subtract to mark three differences in a row, column, or diagonal on the answer board.

- Decide which team will be *X* and which will be *O*. Take turns.
- On each turn, the team picks a number from Sign A and one from Sign B. Then both teams subtract the number on Sign B from the number on Sign A.
- Once both teams agree on the difference, the team whose turn it is finds it on the answer board and writes its *X* or *O* on the number.
- If the team gets a difference that is already marked with an X or O, it loses its turn.
- The first team to write *X* or *O* in three touching differences in a row, column, or diagonal is the winner.
  - These *X*s are in the same row and touch:

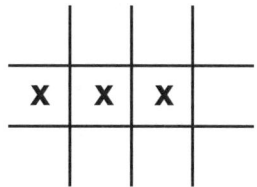

  - These *X*s are in the same row but do not all touch:

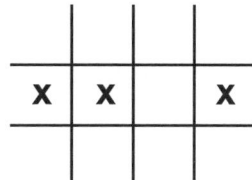

Well Played: Building Mathematical Thinking Through Number Games and Puzzles, Grades K–2 by Linda Dacey, Karen Gartland, and Jayne Bamford Lynch. Copyright © 2016. Stenhouse Publishers.

Word Problem Bingo Cards

Hector had 10 thank-you notes to write. He wrote some of them before dinner. He has 6 more notes to write. How many notes did Hector write before dinner?	There are 5 dogs at the dog park. Five more dogs join them. How many dogs are at the park now?
Tony made 3 paper cranes. He wanted a total of 9 paper cranes, so he could give one to each of his cousins. How many more paper cranes does Tony need to make?	Some students are at the library. Nine of them go to lunch. The other 5 students stay in the library. How many students were in the library before some of them left?
There are 8 lights on in the gym. There are 7 lights off in the gym. How many lights are there in the gym?	The coach bought 6 soccer balls. She bought 7 more baseballs than soccer balls. How many baseballs did the coach buy?
Sam bought 5 small plants and some large ones. He bought 7 plants in all. How many large plants did Sam buy?	Mia had some math problems to solve. She solved 4 of them and had 3 more to do. Altogether, how many math problems did Mia have to solve?

Well Played: Building Mathematical Thinking Through Number Games and Puzzles, Grades K–2 by Linda Dacey, Karen Gartland, and Jayne Bamford Lynch. Copyright © 2016. Stenhouse Publishers.

Word Problem Bingo Cards

There are some sports books on the shelf. The teacher puts 4 more sports books there, and now there are 13 of them. How many sports books were on the shelf before the teacher added some?	Mathew drew a picture with lots of shapes. There were 6 triangles and 5 circles in the picture. In all, how many of the shapes were triangles or circles?
There are 8 fewer boys than girls on the team. There are 11 girls on the team. How many boys are on the team?	There are 6 spoons on the table. There is the same number of forks as spoons on the table. What is the total number of spoons and forks?
Beth has 10 coins. Nine of the coins are dimes, and the rest are pennies. How many pennies does Beth have?	Alesha has 12 shells. She gives 7 of these shells to her sister. How many shells does Alesha have now?
Leah made 16 oatmeal muffins and 8 bran muffins. How many more oatmeal than bran muffins did Leah make?	

Well Played: Building Mathematical Thinking Through Number Games and Puzzles, Grades K–2
by Linda Dacey, Karen Gartland, and Jayne Bamford Lynch. Copyright © 2016. Stenhouse Publishers.

World Problem Bingo Game Board

Word Problem Bingo Directions

Materials Needed
- 1 *Word Problem Bingo* Game Board per team (page A-66)
- 1 deck of *Word Problem Bingo* Cards per group (pages A-64–A-65)
- Optional: 1 *Word Problem Bingo* Directions per group

Directions
Goal: Be the first team to get four answers (or three answers and a free space) in a row, column, or diagonal on your game board.

- Choose a player-leader from one of the teams to both play and lead.
- Write *FREE* in one of the spaces on your team's game board.
- Write the numbers *1–15* randomly in the spaces that are left. Do not write the numbers in order.
- Mix up the cards and place them facedown.
- The player-leader turns over a card, reads the problem, and leaves it faceup.
- Both teams solve the problem and talk about the answer.
- When all players agree, the player-leader records the numerical answer on scrap paper, and each team puts an *X* on the answer on its game board.
- The first team to get four *X*s, or three *X*s and a free space, in a row, column, or diagonal says, "Bingo!"
- Together the teams check the bingo numbers with those the player-leader recorded. If both teams agree that the answers are correct, the team that said, "Bingo," wins.

Well Played: Building Mathematical Thinking Through Number Games and Puzzles, Grades K–2
by Linda Dacey, Karen Gartland, and Jayne Bamford Lynch. Copyright © 2016. Stenhouse Publishers.

Four of a Kind Cards

___ + 9 = 16	13 − 7 = ___
9 + ___ = 16	13 − ___ = 7
16 − ___ = 9	___ + 7 = 13
16 − 9 = ___	7 + ___ = 13

Four of a Kind Cards (continued)

___ − 9 = 5	___ + 7 = 15
___ − 5 = 9	7 + ___ = 15
9 + 5 = ___	15 − ___ = 7
5 + 9 = ___	15 − 7 = ___

Four of a Kind Directions

Materials Needed
› 1 deck of *Four of a Kind* Cards per group (pages A-68–A-69)
› Optional: 1 *Four of a Kind* Directions per group

Directions
Goal: Be the first to get four cards with equations that have the same missing number.

› Four players sit in a circle.
› Shuffle the cards.
› Deal four cards to each player.
› Players look at their own cards and decide on one card they do not want. Each player places this card facedown in front of the player to his or her right. Players put their new cards in their hands.
› Players continue to pass and pick up cards, waiting for all players to pick up before the next pass begins.
› The first player to get four cards with the same missing number says "Four of a kind" and wins.

APPENDIX A-71

It's *Greater* Game Board

Name(s): _____ Date: _____

☐☐ + ☐☐ = _____

☐☐ + ☐☐ = _____

☐☐ − ☐☐ = _____

☐☐ − ☐☐ = _____

Not Using: ☐ ☐ ☐ ☐

You receive 1 point each time your sum or difference is greater.

Score _____

Well Played: Building Mathematical Thinking Through Number Games and Puzzles, Grades K–2
by Linda Dacey, Karen Gartland, and Jayne Bamford Lynch. Copyright © 2016. Stenhouse Publishers.

It's Greater Cards

0	1	2	3
4	5	6	7
8	9		

Well Played: Building Mathematical Thinking Through Number Games and Puzzles, Grades K–2 by Linda Dacey, Karen Gartland, and Jayne Bamford Lynch. Copyright © 2016. Stenhouse Publishers.

It's Greater Directions

Materials Needed
› 1 *It's Greater* Game Board per team (page A-71)
› 1 deck of *It's Greater* Cards, made from 2 copies of page A-72, per group
› Optional: 1 *It's Greater* Directions per group

Directions
Goal: Place digits in the spaces of the game board to create expressions with the greater sums and differences.

› Shuffle the cards and place them facedown in a deck.
› Turn over the top card.
› Each team separately decides in which of the twenty spaces on its board to write the number. Be sure to notice the four *Not Using* squares in which you can write numbers. Once you write a number, it cannot be changed. Then discard this card and turn over the next card.
› Continue playing until you have filled all twenty spaces with a number.
› Teams add and subtract to complete each of their equations.
› Compare your answers to each problem. The team with the greater answer gets 1 point. The team with the most points wins.

Well Played: Building Mathematical Thinking Through Number Games and Puzzles, Grades K–2
by Linda Dacey, Karen Gartland, and Jayne Bamford Lynch. Copyright © 2016. Stenhouse Publishers.

Meet the Rules Game Board

Name(s): _____ Date: _____

1) The value is between 20 and 30. Number sentence:

   _____

2) The value is between 75 and 90. Number sentence:

   _____

3) The value is greater than 300 and less than 350. Number sentence:

   _____

4) The value is between 80 and 100. Number sentence:

   _____

5) The value is greater than 210 and less than 260. Number sentence:

   _____

Well Played: Building Mathematical Thinking Through Number Games and Puzzles, Grades K–2
by Linda Dacey, Karen Gartland, and Jayne Bamford Lynch. Copyright © 2016. Stenhouse Publishers.

Meet the Rules Directions

Materials Needed
- 5 dice per group
- 1 *Meet the Rules* Game Board per team (page A-74)
- Optional: 1 *Meet the Rules* Directions per group

Directions
Goal: Roll dice and try to form expressions to meet the given rules before your opponent does.

- Take turns.
- On its first turn, Team 1 rolls five dice and talks about how to use each of the numbers shown once to meet the first rule. The team can make one-, two-, or three-digit numbers. The team can add or subtract or use both addition and subtraction. The answer must meet the rule.
- If the team can use the numbers to fit the rule, it records the number sentence, has the other team check it, and gives the dice to the other team. On its next turn, Team 1 will try to fit the next rule.
- If the team cannot use the numbers to fit the rule, its turn ends and it gives the dice to Team 2. On its next turn, Team 1 must roll and try to fit this same rule.
- The first team to meet all five rules wins the game.

Make Sense Puzzle A

Name(s): _____   Date: _____

Place each number in only one space so that the puzzle makes sense.

1, 2, 3, 4, 5, 6, 7, 8

☐ + ☐ = 3

☐ + ☐ = 10

☐ − ☐ = 4

☐ − ☐ = 3

Make Sense Puzzle B

Name(s): _____ Date: _____

Place each number in only one space so that the puzzle makes sense.

0, 1, 2, 3, 4, 5, 6, 7, 8, 9

☐ + ☐ = ☐ 4

☐☐ − ☐☐ = ☐ 8

☐ − ☐ = 9

Make Sense Directions

Materials Needed
› 1 *Make Sense* Puzzle (A or B) per team (page A-76 or A-77)
› Optional: 1 *Make Sense* Directions per team

Directions
Goal: Place the given numbers so that all of the equations in the puzzle make sense.

› Work as a team.
› Write each of the given numbers in only one space in the puzzle.
› Check the math. Do all of the equations make sense? If not, try again.
› Check your solution with another team.

Puzzle Answer Key

Chapter 3

Number Jigsaw

Correct arrangements are shown on the reproducibles for the puzzles (pages A-6 and A-7).

Chapter 4

The Number Is/What Number Is?

Correct arrangements are shown on the reproducibles for the puzzles (pages A-25 and A-26), starting at the beginning of each row and proceeding left to right. Any piece may be placed first, and from that piece the order is the same as shown.

Chapter 5

Triangle Totals

Possible solutions are shown below. Students' solutions may be variations of these arrangements, but the same three numbers should be together on a particular side.

Sheet A

```
        1                         4
     6     5                   3     2
   2   4     3              5    1     6
```

Sheet B

```
        4                         5
     9     8                   8     6
   5   7     6              7    4     9
```

Extra puzzles on Sheet B

```
        4                         9
     9     7                   4     5
   6   5     8              8    6     7
```

Puzzle Answer Key (continued)

Chapter 6
Name That Number
Clues A: 5
Clues B: 4
Clues C: 153

Chapter 7
Make Sense
Note that the addends can be reversed.
Puzzle A: $1 + 2 = 3$; $6 + 4 = 10$; $7 - 3 = 4$; $8 - 5 = 3$
Puzzle B: $8 + 6 = 14$; $73 - 45 = 28$; $9 - 0 = 9$

References

Baroody, Arthur. 2006. "Why Children Have Difficulties Mastering the Basic Fact Combinations and How to Help Them." *Teaching Children Mathematics* 13 (1): 22–31.

Booth, Julie L., and Robert S. Siegler. 2006. "Developmental and Individual Differences in Pure Number Estimation." *Developmental Psychology* 41 (6): 189–201.

Bray, Wendy S. 2013. "How to Leverage the Potential of Mathematical Errors." *Teaching Children Mathematics* 19 (7): 424–431.

Brock, Sofia, and Alan Edmunds. 2011. "Parental Involvement: Barriers and Opportunities." *The Journal of Educational Administration and Foundations* 20 (1): 1–23.

Burns, Marilyn. 2007. "Nine Ways to Catch Kids Up." *Educational Leadership* 65 (3): 16–21.

Burton, Meagan. 2010. "Five Strategies for Creating Meaningful Mathematics Experiences in the Primary Years." *Young Children* 65 (6): 92–96.

Cain, Chris R., and Valerie N. Faulkner. 2011. "Teaching Number in the Early Elementary Years." *Teaching Children Mathematics* 18 (5): 288–295.

Chapin, Suzanne H., Catherine O'Connor, and Nancy C. Anderson. 2009. *Classroom Discussions: Using Math Talk to Help Students Learn*. 2nd ed. Sausalito, CA: Math Solutions.

Clements, Douglas H. 1999. "Subitizing: What Is It? Why Teach It?" *Teaching Children Mathematics* 5 (7): 400–405.

Dacey, Linda. 2014. *The How-To Guide for Integrating the Common Core in Mathematics, Grades K–5*. Huntington Beach, CA: Shell Education.

Dacey, Linda, and Anne Collins. 2010a. *Zeroing in on Number and Operations: Key Ideas and Common Misconceptions, Grades 1–2*. Portland, ME: Stenhouse.

———. 2010b. *Zeroing in on Number and Operations: Key Ideas and Common Misperceptions, Grades 3–4*. Portland, ME: Stenhouse.

D'Arcangelo, Maria. 2001. "Wired for Mathematics: A Conversation with Brian Butterworth." *Educational Leadership* 29 (3): 14–19.

Diezmann, Carmel M., Tom Lowrie, and Lindy A. Sugars. 2010. "Primary Students' Success on the Structured Number Line." *Australian Primary Mathematics Classroom* 15 (4): 24–28.

Diezmann, Carmel, and Natalie McCosker. 2011. "Reading Students' Representations." *Teaching Children Mathematics* 18 (October): 162–69.

Eisenhardt, Sara, Molly H. Fisher, Jonathan Thomas, Edna O. Schack, Janet Tasell, and Margaret Yoder. 2014. "Is It Counting, or Is It Adding?" *Teaching Children Mathematics* 20 (8): 498–507.

Erickson, Tim. 1989. *Get It Together: Math Problems for Groups, 4–12*. Berkeley, CA: Lawrence Hall of Science.

Faulkner, Valerie N. 2009. "The Components of Number Sense: An Instructional Model for Teachers." *Teaching Exceptional Children* (41) 5: 24–30.

Ginsburg, Herbert P., and Barbrina Eartle. 2008. "Knowing the Mathematics in Early Childhood Mathematics." In *Contemporary Perspectives on Mathematics in Early Childhood Education*, edited by Olivia N. Saracho and Bernard Spodek, 44–66. New York: Information Age.

Griffin, Sharon. 2004. "Teaching Number Sense." *Educational Leadership* 61 (5): 39–42.

Hillen, Amy F., and Tad Watanabe. 2013. "Mysterious Subtraction." *Teaching Children Mathematics* 20 (5): 294–301.

Jung, Myoungwhon. 2011. "Number Relationships in Preschool." *Teaching Children Mathematics* 19 (11): 551–557.

Kamii, Constance. 2014. "Direct Versus Indirect Teaching of Number Concepts for Ages 4 to 6: The Importance of Thinking." *Young Children* 69 (5): 72–77.

Kazemi, Elham, and Allison Hintz. 2014. *Intentional Talk: How to Structure and Lead Productive Mathematical Discussions*. Portland, ME: Stenhouse.

Kliman, Marlene. 2006. "Math Out of School: Families' Math Game Playing at Home." *School Community Journal* 16 (2): 69–90.

Kling, Gina, and Jennifer M. Bay-Williams. 2014. "Assessing Basic Fact Fluency." *Teaching Children Mathematics* 20 (8): 489–497.

Kohlfeld, Carol. 2009. "Playing Games." *Mathematics Teaching* 215 (September): 14–15.

Koster, Ralph. 2013. *A Theory of Fun for Game Design*. 2nd ed. Sebastopol, CA: O'Reilly Media.

Lan, Yu-Ju, Yao-Ting Sung, Ning-chun Tan, Chiu-Pin Lin, and Kuo-En Chang. 2010. "Mobile-Device-Supported Problem-Based Computational Estimation Instruction for Elementary School Students." *Journal of Educational Technology and Society* 13 (3): 55–69.

López Fernández, Jorge M., and Aileen Velázquez Estrella. 2011. "Contexts for Column Addition and Subtraction." *Teaching Children Mathematics* 17 (9): 540–548.

Marzano, Robert, Debra Pickering, and Jane E. Pollock. 2001. *Classroom Instruction That Works: Research-Based Strategies for Increasing Student Achievement*. Alexandria, VA: Association for Supervision and Curriculum Development.

McNeil, Nicole, and Linda Jarvin. 2007. "When Theories Don't Add Up: Disentangling the Manipulatives Debate." *Theory into Practice* 46 (4): 309–316.

National Council of Teachers of Mathematics. 2000. *Principles and Standards for School Mathematics.* Reston, VA: NCTM.

National Governors Association (NGA) Center for Best Practices and Council of Chief State School Officers (CCSSO). 2010. *Common Core State Standards for Mathematics.* Washington, DC: NGA and CCSSO.

Newman, Rich. 2012. "Goal Setting to Achieve Results." *Leadership* 41 (3): 12–15, 16–18, 38.

Postewait, Kristian B., Michelle R. Adam, and Jeffrey C. Shih. 2003. "Promoting Meaningful Mastery of Addition and Subtraction." *Teaching Children Mathematics* 9 (6): 354–357.

Richardson, Kathy. 2003. *Hiding Assessment.* Assessing Math Concepts, Book 6. Bellingham, WA: Math Perspectives.

Rickard, Caroline. 2013. "Subtraction: Searching for Flexibility." *Mathematics Teaching* 234 (May): 37–39.

Ross, Sharon R. 2002. "Place Value: Problem Solving and Written Assessment." *Teaching Children Mathematics* 8 (7): 419–423.

Sarama, Julie, and Douglas H. Clements. 2009. "Teaching Math in the Primary Grades: The Learning Trajectories Approach." *Young Children* 39 (7): 63–65.

Schiro, Michael S. 2009. *Mega-fun Games and Puzzles for the Elementary Grades: Over 125 Activities That Teach Math Facts and Concepts.* San Francisco: Jossey-Bass.

Stebbins, Leslie. 2003. "Children's Reasoning by Mathematical Induction: Normative Facts, Not Just Casual Facts." *International Journal of Educational Research* 39 (7): 719–742.

Thouless, Helen. 2014. "Whole-Number Place-Value Understanding of Students with Learning Disabilities." PhD diss., University of Washington.

Torbeyns, Joke, Bert De Smedt, Nick Stassens, Pol Chesquière, and Lieven Verschaffel. 2009. "Solving Subtraction Problems by Means of Indirect Addition." *Mathematical Thinking and Learning* 11: 79–91.

Van de Walle, John A., Karen Karp, and Jennifer M. Bay-Williams. 2013. *Elementary and Middle School Mathematics: Teaching Developmentally.* 8th ed. New York: Pearson Education.

Van den Heuvel-Panhuizen, Marja, and Adri Treffers. 2009. "Young Children's Understanding and Application of Subtraction-Related Principles." *Mathematical Thinking and Learning* 11 (1–2): 102–112.

Wedekind, Kassia Omohundro. 2011. *Math Exchanges: Guiding Young Mathematicians in Small-Group Meetings.* Portland, ME: Stenhouse.

White, Jeanne, and Linda Daukas. 2012. "CCSSM: Getting Started in K–Grade 2." *Teaching Children Mathematics* 18 (17): 440–445.